D0408763

THE PETROLEUM INDUSTRY
A NONTECHNICAL GUIDE

THE PETROLEUM INDUSTRY

A NONTECHNICAL GUIDE

CHARLES F. CONAWAY

Copyright © 1999 by
PennWell Publishing Company
1421 South Sheridan/P.O. Box 1260
Tulsa, Oklahoma 74101

Cover Design : Brian Firth
Book Layout and Design : John Potter

Library of Congress–in–Publication Data

Conaway, Charles F.
    The petroleum industry : a nontechnical guide / Charles F. Conaway.
        p. cm.
    Includes index.
    ISBN 0–087814–763–2
        1. Petroleum Engineering. 2. Petrolum industry and trade. I. Title.

TN870 . C58 1999
665.5—dc21
                                                                                                    99–045825

Printed in the United States of America

1 2 3 4 5 03 02 01 00 99

CONTENTS

PREFACE

This book evolved over the last several years as a manual for the Basic Petroleum Technology course I teach around the world for OGCI Training, Inc. Two general precepts underlie the course and book:

- The better we understand the jobs of those around us, the more valuable we are to our companies.

- It's not that hard to get a general understanding of even the most arcane technical jobs.

My goal is to de-mystify the roles of geophysicists, reservoir engineers, etc. so that what they do makes sense to people new to the industry and without technical background. An immediate hurdle is the daunting technical language of these specialists, which makes them seem unapproachable. This book is therefore written in "oilfield," with English explanations when needed. Be prepared to have entirely new images conjured up by words like "dogleg," "mousehole," or "fish".

The book is organized in a natural chronology so that the reader gets an understanding of what precedes and what follows each step. "How the knee bone connects to the thigh bone," if you will. First, we look at how geology formed the earth, the origins of oil and gas, and how it came to be trapped in reservoirs. With the geologic stage set, we then look at the techniques and equipment used to find, drill, produce, process, and market oil and gas. Economic rationales for each step are integrated with the technical issues because this is foremost a business.

My goal has been to produce an easily readable "general" text for the readers to quickly find answers to most questions that come up. If they need to go into more depth, the book will have given them the basic understandings and confidence in the terminology to engage the specialists.

— C. F. C.

A BRIEF HISTORY OF THE PETROLEUM INDUSTRY

The petroleum industry, contrary to most peoples' understanding, did not come about overnight, yet the industry did not take hundreds of years to evolve into what it has become today either. Petroleum based products have been present throughout early European history with uses ranging from healing ointments to combustible weapons of war. Eventually a small oil industry began to develop in Eastern Europe around 1854 that focused primarily on peasant-dug shafts to obtain crude oil. This oil was then refined to abstract kerosene, which was then used in cheaply manufactured lamps. By 1859 a thriving kerosene oil business had been established. Total European crude production in 1859 had been estimated at 36,000 barrels, primarily from Galicia and Romania. More than anything, the Eastern European industry lacked the technology for drilling.

This problem was solved through the determination of one man, George Bissell. Bissell had the inspiration and insight to realize that the same method used for salt "boring" could be used to "bore," or drill, for oil deposits buried beneath the earth. Crude oil, or "rock oil" as it was called, was then a by-product of the salt drilling process and considered a nuisance. Bissell recruited investors in his plan and formed the Pennsylvania Rock Oil Company. Their first objective was to find a promising area of land on which to drill, and then to find someone capable to do the drilling. The man they propositioned was one Edwin L. Drake who happened to meet on of the key investors in the Pennsylvania Rock Oil Company while living in the same hotel. This investor was

James Townsend, a Hew Haven banker, who was impressed with Drake's demeanor and even encouraged him to buy stock in the company. Townsend sent Drake to obtain the title to the prospective oil land on a farm in Titusville, Pennsylvania in 1857. The next step was to begin the process of drilling for the oil. Unfortunately for Drake the local "salt borers" could not be relied upon to stay loyal to the job, and most disappeared or found some other way to get out of the arrangement. It was not until the spring of 1859 that Drake found good fortune in acquiring a driller, a man by the name of William "Uncle Billy" Smith, along with his two sons. Smith was a local blacksmith who made tools for the salt drillers and knew a good amount about what needed to be constructed for the oil drilling operation. The team was fastly draining the monetary resources of the original investors, and most of them stopped their financing of the operation. Townsend was the only investor who still believed in the project, even to the point of paying the venture's bills out of his own pocket. In the same week that Townsend finally lost faith and sent a dispatch to Drake to close up the operation, nothing short of a miracle occurred. On August 27, 1859, the drill dropped into a crevice and then slid another six inches; Drake struck oil.

Unlike the oil gushers, geysers, or "oil fountains" most people commonly associate with striking oil, Drake's discovery had to be hand pumped out of the ground. This event began the mad rush for oil that would engulf the nation in a new era of industrial achievement. The only comparable event as to what took place would be the California Gold Rush. Wells sprung up around the Titusville area like wildfire, but the oil still had to be hand pumped out of the ground, limiting the amount of oil that could be pumped to about 50 bbl/d (barrels per day). All of this changed in 1861 when the world's first flowing well, or geyser, was struck, flowing at about 3000 bbl/d. In 1860, production wavered around 450,000 barrels, then jumped to 3 million barrels in 1862. Unfortunately, production outpaced demand and prices dropped from $10 per barrel to less than 10 cents per barrel within the year, ruining many producers. On the other hand, this swing of prices led to Pennsylvania Oil's dominance in the marketplace by undercutting coal-oils and other illuminants. Demand quickly caught up with supply due to the widespread crossover to Pennsylvania Oil from coal and animal fat products.

The easily acquired fortune and the relatively open market allowed for just about anyone to set up their own production facility. All that was needed was a little start up capital and someone who knew how to drill. These aspects also provided for the frequent fluctuations in the retail price of oil products. Overproduction was a result of a lack of restrictions, for it was not widely understood what exactly went on in these mysterious chambers and veins under the earth. Petroleum geology was a yet unheard of science; for it was still believed that oil discovery was based on luck combined with a natural talent for "sniffing out" oil. It was not until later in the Oil Age that geologists were frequently consulted for information on rock formations, compositions, etc. in relation to the location of oil deposits. This was especially apparent in the case of Patillo Higgins, a one-armed mechanic and lumber merchant from Beaumont, Texas. Higgins believed that oil would be found under a large hill named Spindletop. On the numerous outings Higgins took to the hill, he saw that many of the springs had gas bubbling up into them, and upon poking a stick into the ground near one particular spring he lit the gas that escaped. Every geologist that stopped to survey the area proclaimed that Higgins had no idea what he was talking about and refuted his claim of the presence of oil. Higgins could find no one to support or help him now that most members of respectable society had publicly disavowed him. He would not give up, however, and proceeded to advertise for someone else to drill. He received a response from a Captain Anthony F. Lucas, an experienced prospector of salt in sulfur found in geological structures know as salt domes. The pair then obtained help from the firm of James Guffey and John Galey, who had developed the first major oil field in the mid-continent, in Kansas. This experience paid off with Spindletop, which was a very large salt dome. On January 10, 1901, almost two years after their first attempt, Lucas and Higgins struck what was to be the largest geyser yet. Lucas 1 on Spindletop erupted at 75,000 bbl/d. This sparked the southwestern oil boom, starting with other salt domes across the Gulf Coast of Texas and Louisiana, then spreading into Oklahoma with the vast discoveries at Glenn Pool in 1905, and finally North Texas.

The surplus of oil that was created was only partially due to the increased discoveries of new oil producing locations. "Flush production" was the true father of overproduction. Flush production consisted of racing to produce the most oil out of a well due to the close competition of other producers pumping out of the same location. The lack of geological knowledge combined with the ignorance of exactly how a well worked helped to support this system. Producers had to constantly pump oil out of a well; otherwise their neighboring competitor might produce more oil out of the same deposit. This constant race led to instability in pricing and wasted oil due to excess production.

Transportation of oil was chaotic at best during the beginning stages of the industry, for teamsters would blockade transportation routes to the railroads, charging outrageous rates for through passage. This led to the construction of a wooden pipeline system, which transported oil more efficiently and cheaply to the railroads. By 1866, pipelines were hooked up to most of the wells in the Oil Regions, feeding into a larger pipeline gathering system that connected with the railroads.

Rockefeller's Standardization

John D. Rockefeller was born in 1839 in rural New York State, and lived almost a full century, until 1937. His father moved the family to Ohio when Rockefeller was still a child. While Rockefeller was still a teenager, he aspired to achieve. When the railroad extended its tracks into Cleveland, Rockefeller already owned the largest refinery in the area. Through his own money and that of his investors, Rockefeller threw himself wholeheartedly into the oil business. He constantly expanded his operation, building another refinery and organizing another firm in New York in 1866. In that year, his sales exceeded two million dollars. Rockefeller then proceeded to integrate other aspects of the oil industry such as supply and distribution into the same organization (this would later expand to include production as well). As Rockefeller's organization grew, he became increasingly distraught with the growing oil industry's lack of organization, instability, and poor quality products. There were too many independent producers and refiners competing for the same market causing price instability due to overproduction. There was no set quality control for these products, which led to poor (and even dangerous) product quality. For instance, if one company's kerosene contained too much gasoline, lighting that twilight reading lamp could result in a deadly explosion. Rockefeller set out to eliminate this instability by systematically eliminating the competition, or combining it with his new corporation Standard Oil (which stood for the quality and consistency of a standardized product). Rockefeller would either openly approach a company of interest about selling outright or merging with Rockefeller, wait until the prices of oil plummeted in another industry-wide depression from overproduction, then buy out a company, or drop the prices in an area so that a competing company would be operating at a loss and would be forced to sell. Rockefeller was also known to rely on devious measures to pressure other companies into submission, or to put them out of business. One such scheme associated with Rockefeller was the "South Improvement Company." The South Improvement Company involved a clever scheme of railroad rebates and price drawbacks. The railroads already favored large corporations because of the consistent money they drew in from the continued business by offering rate rebates for continued business. Since Rockefeller's corporation was one of the largest, he greatly benefited from this arrangement; he could sell his products at lower rates than his competitors and still gain a profit. The South Improvement Company went one step further and allowed price "drawbacks." Drawbacks consisted of a percentage of one customer's payment being paid to the larger corporation who meant larger business.

Rockefeller turned Standard Oil into a joint stock company so that he would increase his profit margin without jeopardizing control over the industry. Rockefeller's attempts were extremely successful; by 1879, the war was virtually over. He controlled 90 percent of America's refining capacity." Rockefeller, through Standard Oil, did away with the inefficient and expensive methods in transportation and storage, as well. Wooden barrels were done away with; replaced by steel railway tank cars, horse-drawn tankers for the streets, etc.

Rockefeller was not the only entrepreneur to embark upon the quest for dominance in the oil industry. He faced competition from Russia with the Nobel brothers' (Robert, Ludwig, and Alfred) and Bunge and Palashkovsky (who, with the help of the French family of the Rothschilds, became the second largest producer in Russia). The English were soon to get in on the game, again with the financial help of the Rothschilds, through the efforts of the Samuel brothers (Marcus and Samuel) who were also to innovate a new and safer ocean tanker design that revolutionized the overseas export industry. The Dutch were soon to follow. Aeilko Jans Zijlker started up a refinery under the supervision and support of the Dutch king (William III), called the Royal Dutch company. These foreign competitors increased the efforts of Standard Oil to move into the foreign market; everywhere a competitor appeared Standard Oil would be found. To make matters worse for Rockefeller, none of these foreign companies would concede to merge with his dominant empire.

The advent of electricity threw the oil industry into turmoil yet again in the late 19th century; people now had a much better alternative to petroleum-based illumination at a competitive price. Electricity soon dominated the illumination market within a decade of its inception because of its ease of use and lack of user supervision. Fortunately for the industry, the automobile powered by an internal combustion engine gained credibility in Europe after a Paris-Bordeaux-Paris race in 1895, in which the remarkable speed of 15 miles per hour was achieved. Gasoline, the former by-product of refining kerosene, now had a profitable use. The use of oil products as fuels began at about the same time as the inception of the automobile. Factories, ships, and trains were beginning to transfer over from coal fuel to oil due to its more efficient nature.

References

The Prize: The Epic Quest for Oil, Money, and Power, Daniel Yergin, Joseph Stanislaw, Touchstone Books, 1993

Modern Petroleum: A Basic Primer of the Industry, 3rd Edition, Bill D. Berger, Kenneth E Anderson, PennWell Books, 1992

Petroleum Production in Nontechnical Language, 2nd Edition, Forest Gray, PennWell Books, 1995

HOW THE EARTH WAS FORMED

Origin of the Earth

The mechanism of the earth's origin is still controversial, but in recent years a fair degree of consensus has formed around the Big Bang theory. This holds that all matter in the universe (matter includes everything but space) became compressed into a mass of unimaginable density, then exploded from the stresses created by the compaction. The resulting cosmic dust swirled around the universe. Gravitational forces then drew the dust closer and closer, forming clouds of dust that shrank and grew more dense until they formed the stars and planets.

Structure of the Earth

Gravity continued to shape the planet as it formed. Heavier materials, such as iron and nickel, concentrated in the earth's center, forming the core of the present-day earth (FIGURE 1–1). From analysis of earthquake tremors passing through the earth, seismologists have determined that this core is mostly in a liquid state, but with a solid center.

Surrounding the core is the mantle, composed largely of lighter silicates. Heat and pressure keep it partially melted so that it is not quite a solid and not quite a liquid. It therefore behaves as a plastic, extruding under pressure.

Finally, the solid crust occupies the surface of the sphere. Its materials are the lightest of all. The crust essentially floats on the denser mantle and the mantle floats on the denser core. The force of gravity therefore gravity-segregates the earth, with the densest material in the center, and the density decreasing from the center outward.

To complete the inventory of the earth's parts, there is the hydrosphere (water in the oceans, lakes, rivers, etc.) and the atmosphere. These are also held in place by gravity.

The earth has a superficial resemblance to a tomato. Each has semi-liquid material beneath a solid skin. Their proportions are similar, with the earth being 8000 miles in diameter with a crust less than 30 miles thick.

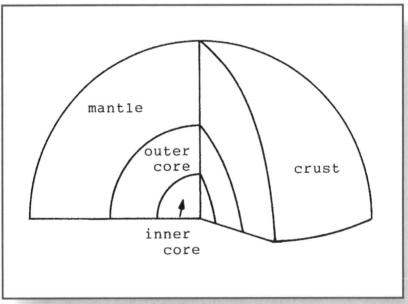

FIG. 1–1   The Structure of the Earth

Tectonics
The Forces that Deform the Earth's Crust

Unlike the quiescent tomato, however, the earth is an extremely dynamic system. Volcanic eruptions and earthquakes signal the powerful forces at work throwing the crust up into mountains and down into basins.

Continental drift

At one point in the earth's past, all the present-day continents were clustered together in a single super-continent (FIGURE 1–2), which has been named Pangaea. Since that time, the continents slowly moved apart until present-day positions were reached. The causes of this continental drift are not entirely understood, but thermal convection currents in the mantle are thought to play a major role.

FIG. 1–2   Location of Present-day Continents at the End of the Permian Period

Convection currents

The earth is a gigantic heat transfer system with extreme heat generated in the core by the weight of the overburden and by radioactive decay. Convection currents in the mantle then move the heat outward to the earth's surface where it is conducted into space.

Convection currents form because hot fluids are lighter than cold fluids. The molecules in fluids (note that "fluid" includes both liquids and gases) are in constant motion and the higher their temperature, the higher their velocity. As the molecules zip about, they collide with one another and bounce apart. Since higher temperature increases velocity, it increases the number of collisions, forcing the molecules further apart and lightening the fluid. Convection currents then form when the hotter (lighter) fluids float upward through colder (heavier) fluids or, conversely, colder fluids drop down through hotter fluids.

For example, housing architects locate heater outlets near the floor so that the heated air rises through and mixes with the cooler air in the room. This results in a somewhat even temperature throughout the room. If the heated air were introduced near the ceiling, it would collect at the top of the room and not mix with the cooler air. The room would then be too hot at head level and too cold at foot level.

The prevailing theory is that the mantle material, called magma, is hot enough to flow like a stiff liquid. Convection currents of hot magma rise to the surface and move laterally beneath the crust until the magma cools and sinks back toward the center of the earth.

Spreading centers

As the currents move laterally under the crust, they impart frictional forces that pull the crust apart. Fresh magma then rises to fill the void, cools, and forms replacement crust. These spreading centers, or rifts, are present in many of the world's ocean basins. For example, an active rift zone called the mid-Atlantic ridge is located in the Atlantic Ocean between North America and Europe and between South America and Africa. It comes onto land in Iceland, causing the fumeroles and geysers.

Subduction zones

Since the spreading centers generate new crust, existing crust must be simultaneously consumed elsewhere in the world. This takes place in subduction zones where two pieces of the earth's crust converge, with one overriding the other (FIGURE 1–3). In this example, the thicker and stiffer continental crust overrides the weaker oceanic crust, forcing it back into the earth where it melts into magma. Compressive forces distort the continental crust and push up mountains, often releasing volcanism. Ocean trenches form at the seam of the two plates where the subducting crust warps downward.

The west coast of South America is an example of an active subduction zone. The Andes Mountains and the offshore Chilean Trench were both created by its action. If two sections of oceanic crust converge, one will override and subduct the other. However, when two sections of continental crust converge, neither is flexible enough to subduct. Instead, they typically lock together in a suture of mountainous masses. The Caucasus Mountains east of the Black Sea are formed by such a suture.

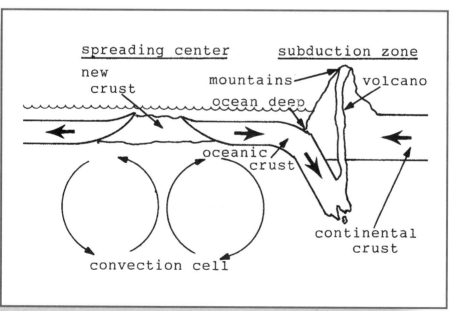

FIG. 1–3 Spreading Center with Oceanic Crust Subducting Beneath Continental Crust

There are also places where two sections of crust grind sideways past one other with essentially no vertical displacement. An example of such transform movement is the San Andreas fault system in California.

Plate tectonics

The movements and distortions of the earth's crust described above are part of the theory of plate tectonics, which has gained wide acceptance in the last 25 years. The earth's crust is visualized as a number of distinct plates (FIGURE 1–4) whose margins are constantly extended by spreading centers and trimmed by subduction. Imbedded in the larger plates, the continents constantly change location, appearing to drift around the earth's surface. Plate tectonics is considered to be the primary mechanism responsible for creating the earth's mountain ranges, ocean trenches, and most of the earthquakes and volcanic activity.

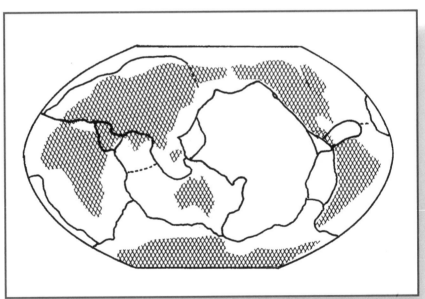

FIG. 1–4   The Major Tectonic Plates in the Earth's Crust

Types of Rock

The earth's crust is composed of rocks, and rocks are composed of minerals—chemical compounds such as quartz (SiO_2) and calcite ($CaCO_3$). When the magma that rises up in the spreading center cools and solidifies, it becomes igneous (fire-formed) rock. Granite and basalt are examples of common igneous rocks.

Most of the rock exposed on the earth's surface, however, is sedimentary rock, which forms when existing rocks break down and are re-deposited. Sandstone and limestone are examples of sedimentary rock.

A third type—metamorphic rock—forms when deeply buried igneous or sedimentary rocks are altered under the intense heat and pressure. Marble and slate are examples of metamorphic rock.

From a petroleum perspective, igneous and metamorphic rocks are unimportant. Oil and gas are formed only in sedimentary rocks and, with few exceptions, are produced only from sedimentary rocks.

Formation of sedimentary rocks

The first step towards creating sedimentary rocks is when tectonic (mountain-building) forces thrust existing rock formations upward as new mountains or highlands. Exposed to the full effects of Mother Nature's great levelers—weathering and erosion—the mountains immediately start to break down.

Weathering. On the top of mountains, it is considerably colder because the insulating blanket of atmosphere is thinner than it is in the valleys and retains less of the earth's heat. Winds gaining altitude to cross over mountains therefore become colder. This lowers their water-carrying capacity, resulting in the high levels of precipitation typical of high altitudes. The plentiful water available on mountains accelerates weathering processes. Frost wedging—the single most effective mechanism in the breakdown of mountains—begins with ground waters penetrating minute fractures in the rock. The weather then turns cold and freezes the water in the fractures. As the water changes to ice, it expands slightly, but with a force great enough to shatter the rock (Figure 1–5).

This is the same process that forms potholes in paved highways. The weather most destructive to both roads and mountains is daily thawing and nightly freezing where there is ample moisture available.Water is also essential to the chemical weathering of rocks. For example, carbon dioxide absorbed by rain as it falls through the atmosphere often creates slight acidity in surface waters. Over time, limestone rock in contact with these waters can be dissolved away by this dilute carbonic acid. Enormous limestone caverns around the globe testify to the effectiveness of this process.

This mechanism also can convert impervious limestones into porous and permeable rocks suitable for petroleum reservoirs. The pencil-lead sized solution channels, or vugs, in a limestone reservoir were formed in the same way as underground caves. They just didn't have time to get as big. Acidic ground waters are also effective in weathering the igneous mineral Feldspar into clay.

Erosion.   The high rainfall on mountain tops, combined with the steep slopes, sends streams coursing downward laden with the rubble from weathering. As the rubble tumbles in the fast-moving waters, it breaks into smaller and smaller pieces. Chunks of rock also impact the streambed, gouging out additional pieces of rock as the stream cuts its channel deeper.

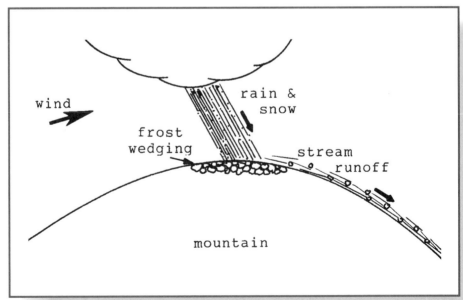

FIG. 1–5   Mountain Undergoing Weathering and Erosion

Erosion is nature's leveler. It transports weathered material from higher altitudes and deposits it at lower altitudes. If it were not for the continual up-welling of new mountains, weathering and erosion would eventually make the entire world level and featureless.

Deposition. A stream tumbling down a mountain transports a heavy load of solid material. However, as it approaches the lowlands and its velocity decreases, the larger particles start to drop out of suspension, with progressively smaller particles following. This is clastic deposition, which means the deposition of material consisting of fragments or grains.

Figure 1–6 shows a river entering the ocean. As the current spreads out and slows, it deposits coarse gravel close to shore, the smaller sand particles a little farther out, and the low density, clay particles in deeper water.

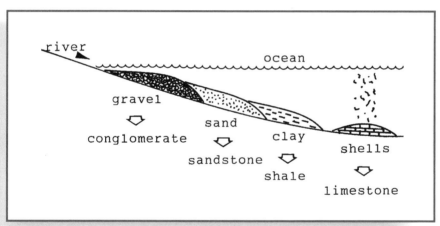

FIG. 1–6 Sediment Deposits and Resulting Sedimentary Rocks

Another form of clastic deposition takes place in the open sea, but is not associated with runoff from a landmass. A host of small planktonic (free-floating) organisms live in the oxygenated and sunlit surface waters. Some have shells that, when the organisms die, sink to the bottom and accumulate on the seabed.

In addition to clastic deposition, chemical deposition also takes place. Evaporation in lagoons isolated behind reefs or bars can increase concentrations of dissolved calcium carbonate, causing it to precipitate out as lime mud (FIGURE 1–7).

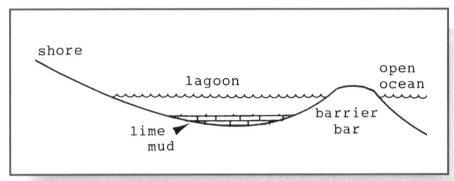

FIG. 1–7   Lime Mud Deposition in Lagoon

Lithification. Lithification is the conversion of unconsolidated sediments into sedimentary rock by compaction and cementation. As deposition proceeds, earlier sediments are buried deeper and deeper. This increasing overburden squeezes out water, compacting the sediments until individual particles come into contact. After the geometry is fixed by compaction, cementation by calcium carbonate or silica precipitating from circulating waters may glue the rock grains together.

Lithification of sediments yields the following sedimentary rocks.

- Gravel becomes conglomerate, which is sometimes a petroleum reservoir rock.

- Sand becomes sandstone, which is the most common petroleum reservoir rock.

- Clay becomes shale, the most common sedimentary rock.

- Lime mud becomes a non-porous, usually gray to black limestone.

- The microscopic carbonaceous shells become chalks, a fine-grained, white limestone. The "White Cliffs of Dover" outcropping on the English coast is a chalk formation as is the reservoir rock for the giant Ekofisk Field in the North Sea.

Rock formations

Sedimentary rocks usually occur as formations—relatively thin layers extending over broad areas. Geologists assign names to the individual formations, e.g., Woodbine Sandstone, Atoka Limestone, etc. Formations are the basic unit used to describe and map sedimentary geology.

Figure 1–6 illustrates how particles of similar size and density are deposited roughly the same distance from shore. Over time, however, this mechanism would create vertical buildups of sediments—not the horizontal sheet-like formations that are actually found. A second mechanism must therefore be involved.

Just as there is horizontal movement of the earth's crust with the various plates jostling one another, there is also continuous vertical movement. For example, mountain-building raises some areas while the weight of thick sediments lowers others. This movement is subtle, essentially unnoticeable, but with the eternity of geologic time its effects become very significant.

Figure 1–8 illustrates how a submergent coastline migrates landward with the offshore deposition migrating in the same direction. As a result, the sands are deposited not in a vertical pile, but as a continuous horizontal layer typical of formations. The clays and gravels are similarly deposited in blanket configurations, forming layers on top and beneath the sand.

Emergent coastlines (lowering water level) also generate blanket-type deposits, but the deposits grow in a seaward direction. Frequently a deposi-

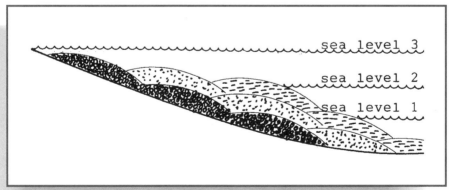

FIG. 1–8 Submergent Shoreline Generating Blanket-type Deposits— the Precursors to Rock Formations.

tional event will span a number of reversals from emergence to submergence and back again. This results in repeated lithological sequences. For example, wells drilled offshore (Nigeria or the U.S. Gulf coast and elsewhere) typically have multiple sandstone reservoirs separated by shale zones. These sand-shale sequences were formed by repeated emergent-submergent cycles.

Geologic Structure

In geology, the word "structure" refers to the distortion of rock formations by the earth's forces.

Folding

Near subduction zones, the compressive force of plate movement buckles the crust into multiple folds arranged perpendicular to the direction of movement. Although the distortion may fracture surface rocks, the deeper seated rocks are hot and therefore quite plastic. As a result, petroleum geologists routinely encounter smooth, unbroken folds.

Folded formations contain two structural features. The bottoms of the folds (concave upward) are called synclines whereas the tops of the folds (concave downward) are called anticlines. Anticlines, also called domes or highs, are of particular interest in petroleum exploration because they can provide traps for petroleum.

Faulting

Faults are structures formed when subsurface forces fracture crustal rocks and slippage occurs along the fracture plane. Like anticlines, faults are frequently involved in trapping petroleum (FIGURE 1–9).

Normal faults occur with the crust in tension. When it fractures, slippage along the fault plane lengthens the crust. An example would be growth faults that form around the perimeter of depositional basins. As the weight of sediment build-up bends the crust downward, generating tensional stress, movement occurs along the fault planes to relieve the stress.

Faults resulting from compressional forces are reverse faults. They shorten the crust by riding up on the offsetting fault blocks.

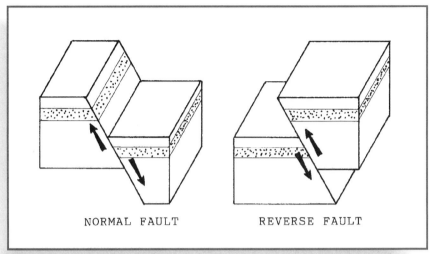

NORMAL FAULT   REVERSE FAULT

Fig. 1–9   Spreading Center with Oceanic Crust Subducting Beneath Continental Crust

A transverse or slip fault occurs when lateral crustal stresses cause horizontal movement along fault planes. An example is the enormous San Andreas fault system in California where two major crustal plates move laterally to each other. It is not unusual for normal and reverse faults to also have transverse movement involved.

Geology of Common Land Forms

The following are examples of some of the common land forms in the world with explanations of their geology.

Ridge and valley areas

Long narrow ridges separating broad valleys can result from folding and erosion. A harder, more competent rock formation erodes slower, so its outcrops form the ridges. Softer formations erode more rapidly, so their outcrops form the valleys.

The folds shown in Figure 1–10 include both a syncline and an anticline. These structural features are caused by horizontal compressive forces. Only the roots of the anticline remain after erosion. The harder rock emerges from the top of one ridge and returns to earth at the next, with the intervening loop no longer present.

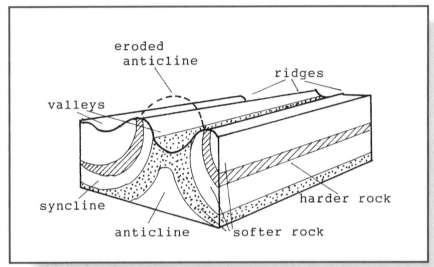

FIG. 1–10   Ridge and Valley Erosional Surface

Glaciated regions

The temperature of the earth's surface has varied significantly over geologic history. During periods of low temperature, snow accumulated, turned to ice, and formed glaciers. Essentially rivers of ice, the glaciers swept down from the north, covering much of the land surface of the Northern Hemisphere.

The movement of the glaciers gouged out long, narrow valleys in the underlying rock. The fjords of Scandinavia, the lochs of Scotland, and the finger lakes in the United States are examples. U-shaped cross-sections and hanging waterfalls are typical of glacial valleys (FIGURE 1–11).

At the end of a glacial period, the ice melted and retreated northward. Glacial till composed of rock fragments, boulders, and rock dust embedded in the glacier dropped out and was deposited as moraines. Today, glaciated areas can often be identified by an abundance of rounded boulders. In many cases, the farmers have built fences with them.

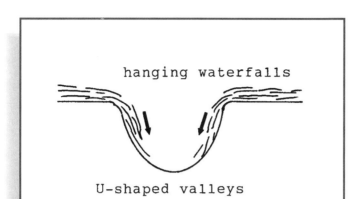

Fig. 1–11   Cross-section of Glacial Valleys

Mountains—new and old

Mountains that were thrown up recently have sharp, jagged profiles. Examples are the Alps and the Canadian Rockies. Older mountains that have been subject to erosion for a considerable time are more rounded. Examples are the Urals and Appalachians.

Mesas

Mesas are flat-topped elevated erosional features. The bedding plane of the underlying sedimentary rock is horizontal. A particularly competent formation—often a well-cemented sandstone—forms a cap rock which resists erosional forces, maintaining the mesa's elevation (FIGURE 1–12).

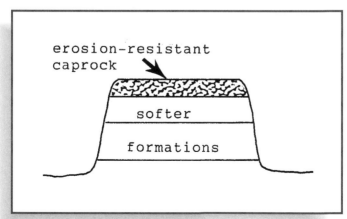

Fig. 1–12   Structure of Mesas

Rivers—new and old

New rivers can be characterized as having a steep gradient, and are associated with abrupt topographic features. As they plunge rapidly downward, their course is straight and they cut V-shaped valleys (FIGURE 1–13).

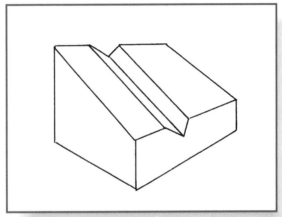

FIG. 1–13   Steep Gradient Waterwater Course

Mature rivers are typical of plains areas where erosion has removed most topography. The river's gradient is gentle, its velocity low and its course has many bends, called meanders (FIGURE 1–14).

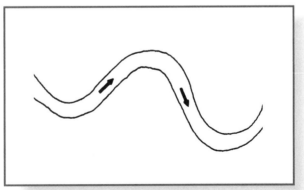

FIG. 1–14   Meandering water Course

Oxbow lakes

Numerous oxbow lakes are found along the course of a mature river. They represent vestigial meanders that have been cut off from the stream. Figure 1–15 illustrates how oxbow lakes are formed.

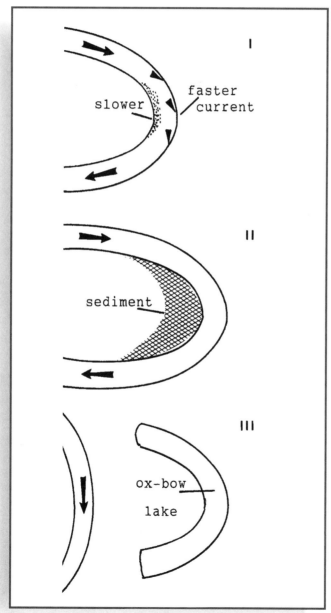

I

slower   faster / current

II

sediment

III

ox-bow lake

Fig. 1–15   Evolution of an Oxbow Lake

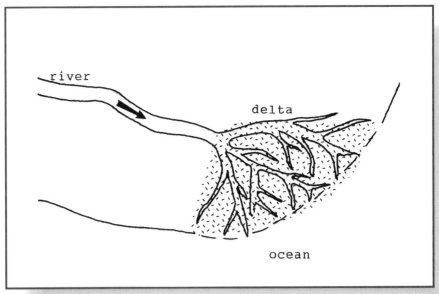

FIG. 1–16   River Delta

River deltas

When a river enters an ocean, its waters spread out and slow down. Suspended solids drop out and accumulate on the seafloor, forming a fan-shaped delta.

Delta deposits are of great interest to petroleum geologists because many of the world's most prolific oil and gas fields are in deltaic sandstones. Examples are the Niger River and Mississippi River deltas (FIGURE 1–16).

The sequence of events in forming a delta is for the river to enter the ocean through only one or two passes at a time. The sediments accumulate at the mouths of these active passes, raising their water level. Periodically, during high water conditions, the current jumps the bank of an active pass and cuts a new pass to the ocean, leaving the original pass dry. Over time this creates the triangular, or fan shape of deltas.

PETROLEUM ORIGINS AND ACCUMULATION 2

Origin of Petroleum

There is still some disagreement on the origins of petroleum and the mechanisms by which it is concentrated into commercially sized accumulations. Probably the most controversial issue is whether petroleum has an inorganic or an organic origin. That is, whether it is derived from chemical reactions between minerals or from tissue created by living organisms.

The inorganic thesis has been particularly advocated by geologists from eastern European countries, but has other adherents as well. For example, the Swedish government recently drilled several unsuccessful exploratory wells in the granite that forms the bulk of their country. The wells were in the impact crater of an ancient meteor. The reasoning was that the meteor would have shattered the granite, providing flow channels to the wellbore for any oil and gas it contained.

Methane (natural gas) can be generated in the laboratory by applying heat and pressure to naturally occurring minerals, so its likely that some petroleum has an inorganic origin. The consensus of opinion is, however, that the vast majority of oil and gas is derived from organic material (FIGURE 2–1).

Which organic source?

It's easy to imagine the great dinosaurs and huge prehistoric fishes being pulverized and rendered into petroleum. Unfortunately, it turns out that there have not been nearly enough of these "recognizable" animals alive in the history of the world to account for the incredible quantity of oil and gas that has already been found. Another explanation is needed.

In fact, the only organisms with enough mass to possibly account for so much oil and gas are the many forms of microscopic plankton and algae living in the oxygenated near–surface waters of the oceans. The highly mineralized polar waters support these tiny crustaceans, worms, diatoms, etc. in such abundance that they cloud the water. As the base of the food chain, they are preyed upon by macroscopic plankton, which then support the entire community of fish, seabirds, and marine mammals.

FIG. 2–1   Some scientists believe petroleum formation began millions of years ago, when tiny marine creatures abounded in the seas (courtesy API)

Two conditions are necessary for plankton and algae to be the source of petroleum. First, wholesale death is necessary to provide sufficient volume for commercial development. Second, rapid burial of the dying organisms is necessary to prevent bacteria from consuming them.

A present–day example of a good environment for oil formation is where the Straits of Bosporus enter the Black Sea near Istanbul, Turkey. Currents from the Mediterranean rush through the strait into the Black Sea, and immediately plunge to great depths. The waters are rich in micro–organisms that undergo wholesale death as they are carried downward into non–oxygenated waters. The dead organisms sink to the bottom and are quickly buried by the rapid clay deposition in the area. This protects them from bacterial action (FIGURE 2–2).

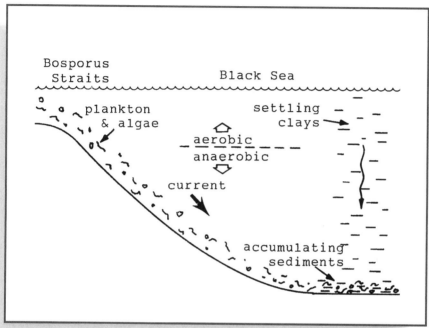

FIG. 2–2   Deposition and Burial of Organics

When organic–laden clays are lithified into shale, they typically are black in color. Black marine (oceanic) shale is therefore considered to be the source rock for most oil.

The presence of adequate source rocks—usually black marine shales— is an important consideration in petroleum exploration. Source rocks are a particular focus when exploration geologists are first evaluating a new basin. Once oil has been discovered anywhere in the basin, however, identifying source rocks is no longer of concern because they obviously exist.

Vegetation as a source of natural gas

It's thought that virtually all "black oil"—meaning oil with a strong color (which excludes gas condensate)—is formed as above from plankton and algae. All the gas associated with black oil (gas in solution in the oil and gas-cap gas) has a similar origin, as do most gas fields. However, some natural gas originates from decaying vegetable matter that was laid down in shallow, freshwater environments. This is known as microbial gas because it is generated by bacterial action. The gas in the southern North Sea fields, for example, is thought to have formed from vegetation associated with an ancient tropical river delta in the area. The growing coalbed methane industry also produces microbial gas by drilling to deep coalbeds and producing the coal gas.

Chemistry of Petroleum

Hydrocarbon molecules

Oil and natural gas are made up of hydrocarbon molecules. Molecules, the basic building blocks of nature, are composed of from one to thousands of atoms. For example, the water molecule has three atoms—two hydrogen and one oxygen. The atoms are held in a three–dimensional matrix by valence bonds—an electrical attraction.

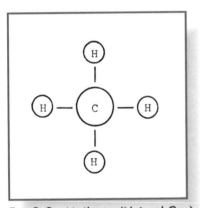

Hydrocarbon molecules occur only in living things—plants and animals—and consist of hydrogen and carbon atoms. The carbon atom has a valence of four while the hydrogen atom has a valence of one. It therefore takes four hydrogen atoms to satisfy one carbon atom. This is the description of the smallest hydrocarbon molecule—methane, also called natural gas. Methane is, by far, the most common hydrocarbon molecule (FIGURE 2–3).

FIG. 2–3 Methane (Natural Gas) Molecule

The next larger hydrocarbon molecule is ethane, which has two carbon atoms and six hydrogen atoms. In Figure 2–4, note the valence bond connecting the carbons which leaves three open positions on each carbon to be occupied by hydrogens.

Ethane is nearly as volatile as methane and so is often included with the methane in residue gas (marketable gas remaining after extracting the heavier hydrocarbons, water, and other impurities). The ethane enhances the heat content of the residue gas stream, increasing its value. When petrochemical plants are in the vicinity, however, it's often profitable to install cryogenic (extreme low–temperature) gas processing to separate out the ethane and sell it as a petrochemical feedstock.

Figure 2–4 shows three paraffinic or "straight–chain" molecules. Ethane has two carbons, propane has three, and butane has four. The number of hydrogens in a given paraffinic molecule can be derived from the formula $H = 2n + 2$, where n equals the number of carbon atoms in the molecule. Hence the number of hydrogens in propane is $(2 \times 3) + 2$, or 8.

Propane and butane are gaseous at atmospheric conditions, but readily liquefy under moderate pressure. They are marketed in liquid form as bottled gas and as chemical feedstocks.

Just as there are paraffinic hydrocarbon molecules with one, two, three, and four carbons, there are also molecules with five, ten, and many more carbons. In addition to paraffins, crude oil contains ring, or cyclo, compounds. For example, Figure 2–5 shows the five–carbon naphthenic molecule cyclopentane and the six–carbon aromatic molecule benzene.

The smaller the hydrocarbon molecule, the more volatile it is. Methane, ethane, propane, and butane are defined as petroleum gases because they are in a gaseous state at atmospheric conditions. Hydrocarbons from C_5 and higher molecules are defined as petroleum liquids because they are liquid at atmospheric conditions. In general, the larger the size of the hydrocarbon molecule (number of carbons), the greater its viscosity, the lower its volatility and the darker its color.

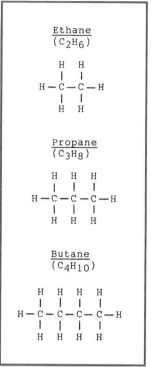

FIG. 2–4   Paraffinic Molecules

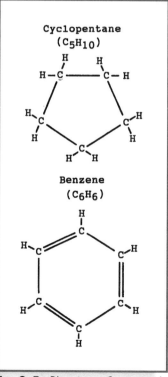

FIG. 2–5   Ring-type Componds in Crude Oil

Crude oils are mixtures of many sizes and types of hydrocarbon molecules, and the character of each crude is determined by its relative proportions of the different molecules. As a result, crude oils vary greatly, with no two being identical. At the light extreme are colorless gas condensates that, if uncontained, completely evaporate in minutes. At the heavy extreme are highly viscous tars that never evaporate and require heat before they can be poured. This extreme variation results from the wide variety of conditions existing as the crude was generated.

Generation of Oil

The virgin oil in freshly deposited organic material is chemically quite different from the crude oil that is later produced. This transformation, called diagenesis, is caused by the high temperatures encountered by the organics as they are buried deeper and deeper in a sedimentary basin.

The deeper the burial, the higher the temperatures and the greater the diagenic effects. Longer exposure to the heat also increases the effect. Before a wildcat is drilled in an unexplored area, geochemists often research its geologic history to try and predict the degree of diagenesis that has taken place. This can suggest that the area will be either oil or gas prone, or that neither will be present. If oil–prone, it may indicate the probable quality of the oil. This information may determine whether or not the wildcat is drilled.

Organics that were not buried very deeply or for very long yield tarlike heavy oils. The deeper and longer the burial, the greater the maturation of the oil. That is, it becomes progressively lighter oil, which means it has lower viscosity, greater volatility (because it includes more small molecules), and is lighter in color. This occurs because the increased heat will "crack," or split apart, the larger liquid molecules into smaller gas molecules, creating a lighter crude. When burial depth is greater than 18,000 ft. where temperatures exceed 300°F, its unlikely that any black oil survives—only gas will be present.

Reservoir Rock Properties

Many people unfamiliar with the petroleum industry visualize oil existing in "underground lakes"—essentially buried stock tanks (FIGURE 2–6). If this were true, finding and producing oil and gas would be extremely simple and inexpensive. The reality is, however, that oil and gas reservoirs are in "solid rock," with the hydrocarbons occupying only the tiny pores and fractures within the rock.

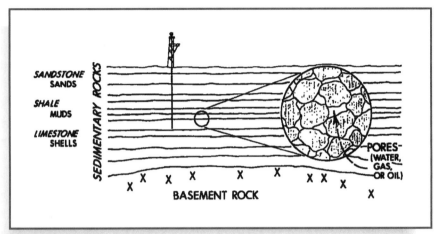

FIG. 2–6   Crude is Not Found in Underground Lakes, But in Solid Rock

To get from the reservoir into the wellbore, oil and gas must negotiate tortuous passageways through the rock, severely restricting the well's production rate. Instead of the few days it would take to pump out an underground lake, it takes many years to deplete an actual reservoir. This seriously impacts profitability. The practice of reservoir engineering was developed to deal with these unique problems.

Porosity. To function as a reservoir rock, a rock must have some open space within its structure to contain the oil or gas. This is called porosity, and is expressed as a percent of the open space to the overall rock volume. That is, a rock with 25% porosity is composed of 75% rock and 25% open space. Φ, the Greek letter Phi, is the symbol used to denote porosity.

There are three types of porosity, with inter–granular porosity the most common. Inter–granular porosity is the open space between the grains or fragments of clastic rocks such as sandstones, conglomerates and clastic limestones.

When identical spheres are stacked as in Figure 2–7, the porosity is 48%, regardless of the size of the spheres. Of course nature would not stack the grains so neatly directly on top of one another, nor would they all be identical. The 48% is therefore a theoretical maximum.

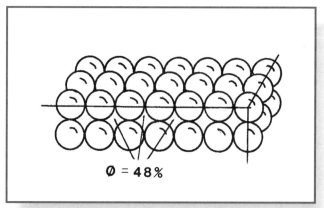

Ø = 48%

FIG. 2–7   Stacked, Same-size Spheres

It is not unusual, however, to encounter sandstones with porosity as high as 25%. This requires that the sand grains be relatively well sorted (close to the same size) and fairly well rounded (not angular). These are characteristics of beach–type deposits.

Only a few sand grains appear to touch in Figure 2–8, suggesting the grains are not packed closely together. However, this cross–sectional view shows only a single plane, so many hidden contact points exist on either side of that plane.

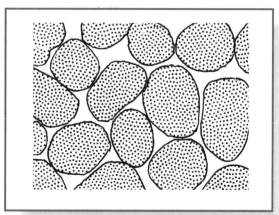

FIG. 2–8   Sandstone with Rounded,
Well-sorted Grains

The average grain size is another important characteristic of clastic reservoirs, but does not directly effect porosity. That is, the grains shown in Figure 2–8 could be large or small, but the porosity would remain the same.

The degree of grain–size sorting, however, has a profound effect on porosity. Figure 2–9 shows similar large grains as were present in Figure 2–8, but with smaller grains added between them, plugging up the pore spaces. This obviously can drastically reduce porosity.

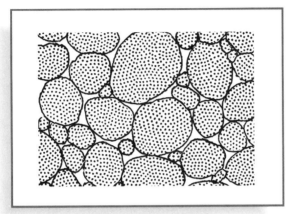

FIG. 2–9   Poor Sorting Can Sharply
Reduce Porosity

The second significant porosity type is vugular porosity. This occurs in limestones that after lithification were lifted up above ocean level and exposed to fresh surface waters. Acidic ground water percolates down through the limestone, dissolving networks of tiny, interconnected channels, or vugs (FIGURE 2–10).

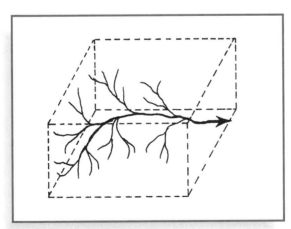

FIG. 2–10   Vugular Porosity

The vugs are normally quite narrow, occupying very little space, so vugular porosity is typically quite low. Nevertheless, some of the largest and most prolific reservoirs in the world are in vugular limestone. Although the porosity may be less than 5%, it can be compensated for by a thicker formation.

The third porosity type is fracture porosity. Fractures occur when hard, competent rock formations are subjected to wrenching tectonic stresses. Porosity exists when the rock remains in stress, maintaining the fractures' offset; that is, not allowing the fracture to heal.

Fracture porosity typically is even lower than vugular porosity. It rarely can be commercial by itself, but often enhances intergranular or vugular porosity. In this situation, the fracture porosity is called secondary porosity while the intergranular or vugular porosity is the primary porosity.

Permeability. A reservoir rock's porosity must be inter–connected for the oil and gas to flow through the formation to the wellbore. This "connectiveness" is called permeability and is measured in darcys. Vugular porosity has high permeability because the vugs are actual flow channels. Fractures are also excellent flow channels, so fracture porosity also has high permeability.

The permeability of inter–granular porosity is highly variable. For example, sandstones recently deposited on continental margins are often loosely consolidated and without cementation. Permeability is therefore high.

Conversely, the geologically older sandstones typical of onshore formations have been subjected to more tectonic stress and cementation. Cementation has little effect on porosity, but can severely restrict permeability (FIGURE 2–11).

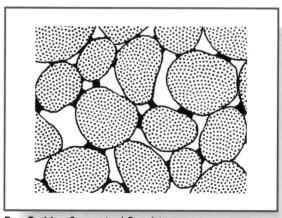

FIG. 2–11 Cemented Sandstone

Fluid Saturations. Because the sediments were laid down in oceans, they were originally surrounded by sea water. As a result, salt water occupied the porosity when the sediment became lithified. It remains that way today unless some of the salt water was later displaced by oil, gas, or fresh water. These displacement mechanisms and their effectiveness are of great interest to reservoir engineers because they determine the profitability of the reservoir.

When migrating into a reservoir rock, oil and gas do not cleanly flush all the saltwater from the rock's pores. The displacement process is slow, requiring geologic time, so in most reservoirs the process is not complete. Even in completely displaced reservoirs, some water remains behind, clinging to the surfaces of the rock grains. Saltwater therefore always occupies some of the porosity, reducing the volume available for oil or gas.

Relative permeability. Oil and gas do not go into solution with saltwater in the reservoir. Instead, they remain separate from the water, forming discrete phases. Figure 2–12 illustrates how the water and oil phases co–exist in the pores when there is a low oil saturation, perhaps 25%, and a high water saturation. In this case, the oil's surface tension draws it into discrete globules located in the main pores (a low gas saturation would behave similarly, creating discrete gas bubbles). If a well were completed in the zone, water would flow freely because it is a continuous phase connecting throughout the porosity matrix. The oil, however, is in discrete globules—not continuous—and so it would not flow. This reservoir would produce 100% water.

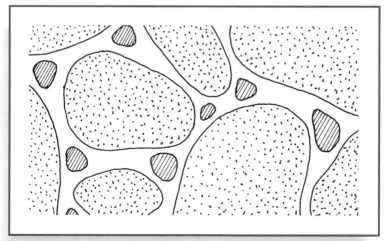

FIG. 2–12   Sandstone With Low Oil and High Water Saturations. Oil is Non–continuous.

Figure 2–13 illustrates the fluid patterns when oil has completely displaced all movable saltwater in the pore matrix. The oil saturation (So) is very high, probably in excess of 80%, and water saturation (Sw) is less than 20%. The water remaining, called the irreducible minimum saturation, is immovable, bound to the sand grains by capillary forces. Conversely, the So is so high that the oil phase connects between pores, permitting flow when a pressure differential is applied. Since the oil phase is continuous between pores and the water phase is not continuous, this reservoir would produce 100% oil.

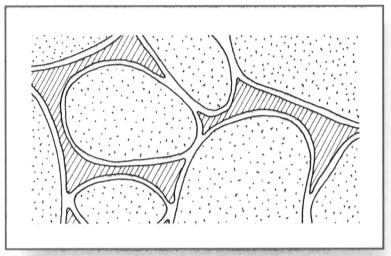

FIG. 2–13 Sandstone with Oil Saturation High Enough to be Continuous Between Pores and Therefore Producible.

Many reservoirs produce both oil and water. For this to happen, water must exceed the irreducible water saturation and oil must exceed the residual oil saturation—the minimum So where the oil is continuous. Every formation is different, but a rough estimate is that dual phase flow can be expected whenever both So and Sw exceed 25%.

The above discussion outlines the relative permeability effects that various saturations have on what fluid or fluids the reservoir produces. When a prospective formation is penetrated by the drill bit, it therefore becomes a pressing matter to find out the saturations of oil, water, and gas. This information is typically obtained from the open–hole electric logs (see Chapter 6) that are run after the hole reaches total depth.

Migration and Entrapment

The isolated droplets of petroleum embedded in black marine shales or other source beds are so broadly and thinly dispersed that there is no economic way to recover them. To have value, they must be concentrated into huge accumulations that can justify the costs of drilling and production.

Expulsion. The first step toward concentration is for the droplets to escape from the confining shale. Since the shale is composed of clay platelets that seal off any passages, it is impermeable under normal conditions. However, normal conditions are far exceeded as the shale is buried deeper and the weight of the overburden increases. This pressure compacts and fractures the shale, squeezing out water and hydrocarbons (FIGURE 2–14).

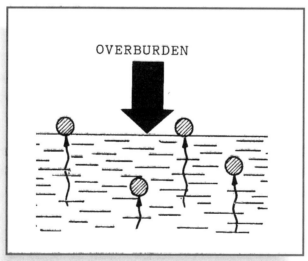

OVERBURDEN

FIG. 2–14 Oil Droplets Expelled from Shale

Vertical migration. When the oil droplets are squeezed out of the shale and into a porous, permeable sandstone saturated with saltwater, they have both the space and the energy to move. The direction of movement, or migration, will initially be upward because oil is lighter than water. The actual route taken by a given droplet may be quite tortuous as it negotiates the twists and turns of the porosity matrix; but given geologic time, it can cover significant distances (FIGURE 2–15).

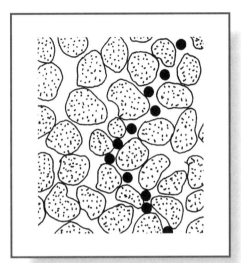

FIG. 2–15   The Torturous Upward
Movement of Oil Drops

Horizontal migration and trapping. In Figure 2–16, the sandstone is over-lain by another shale, which prevents further vertical movement. Instead, the droplets are diverted to the right by the slope of the base of the shale. They continue to migrate diagonally upward until the slope of the shale reverses, forming an anticlinal structure. The oil then collects in this anticlinal trap. The overlying shale is the cap rock for the trap. Shales are the most common cap rocks, but any impermeable rock will serve. The sandstone is the reservoir rock, which must be porous and permeable.

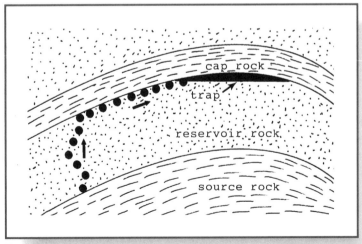

FIG. 2–16   Vertical and Horizontal Movement of Oil into Trap

The source rock can be adjacent to the reservoir rock as shown in Figure 2–16, or can be miles away. There is evidence that gas with its low viscosity can migrate laterally for hundreds of miles. Because of its higher viscosity, oil probably does not migrate that far, but there are documented cases where oil has migrated as far as 50 miles.

The oil is trapped in the top of the structure by gravity segregation—that is, because it is lighter than water. If the trap held both oil and gas, the gas would be on top of the oil as a gas cap—again, by gravity segregation. The oil–water and gas–oil contacts are, with few exceptions, horizontal (FIGURE 2–17).

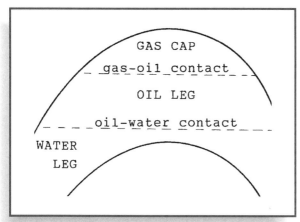

FIG. 2–17   Reservoir with Gas Cap

For an effective trap to exist, there must be closure. One way to visualize closure is to think of a soup bowl. If one side were cut out of it, it would hold no liquid. To be effective, it must contain the liquid on all sides—a full 360°. A trap is like an upside–down soup bowl. It must contain the oil and gas on all sides or it will leak off and continue its migration (FIGURE 2–18).

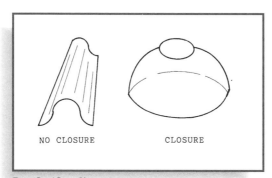

FIG. 2–18   Closure

Types of Traps. There are many different trap configurations. They are classified as structural traps, stratigraphic traps, or traps that are a combination of both.

Structural traps. Structural traps are created by deformation of the rocks. Anticlinal, or domal, traps discussed earlier are one example. They are caused when horizontal compressional forces deform the rock into folds.

Another very common structural trap is the fault trap. The vertical displacement of the fault interrupts the continuity of the reservoir rock, moving a section of impermeable rock into position where it seals off the reservoir (FIGURE 2–19).

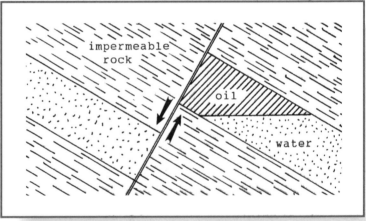

FIG. **2–19** Fault Trap

In salt dome provinces such as the Texas Gulf Coast, numerous opportunities for structural traps exist. The weight of overlying sediments extrudes deep–seated salt deposits into upward thrusting domes. These forces create numerous faults in the surrounding sediments, which often form fault traps. In addition, anticlinal traps are pushed up on the crest of the dome and oil is trapped against the impenetrable salt along the flanks of the dome (FIGURE 2–20).

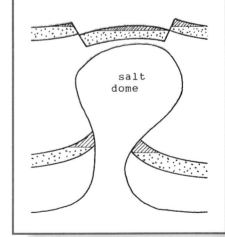

FIG. **2–20** Salt Dome Trap

Stratigraphic traps. Stratigraphic traps are formed by depositional conditions rather than deformation of the rocks. For example, when an inclined bed of permeable sandstone changes into shale up–dip, the impermeable shale functions as a cap rock. Such up–dip pinchouts are quite common since they reflect the normal transition from sand to clay deposition that takes places at greater distances from shore (FIGURE 2–21).

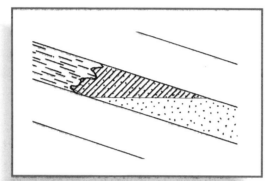

FIG. 2–21   Up–Dip Pinchout

Coral reefs, or bioherms, that become buried under sediment can become prolific stratigraphic traps. Reefal porosity is the space once occupied by the living animal, so it can be very high, as can permeability. Although reef traps typically cover a limited surface area, they often are quite thick. This geometry and the high quality of the porosity and permeability often result in high well production rates (FIGURE 2–22).

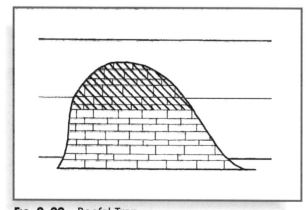

FIG. 2–22   Reefal Trap

Some other common types of stratigraphic traps are formed from

- Channel sands deposited in sinuous bands parallel to river channels. When buried by subsequent deposition, they can become high quality reservoir rocks.

- Offshore barrier bars. When these long narrow sand deposits are buried by subsequent shale deposition, they can become high quality reservoir rocks.

Combination traps. Combination traps incorporate both deformational and depositional mechanisms. An example is an angular unconformity (FIGURE 2-23), which requires several steps in its formation.

1. The reservoir rock is deposited and lithified in an ocean basin.

2. The formations are pushed above sea level and tilted.

3. A horizontal erosional surface develops on the tilted formations.

4. The erosional surface sinks below the ocean's surface and horizontal shale deposition covers the exposed ends of the formations

5. The shale lithifies and becomes the cap rock for the tilted reservoir rock.

Despite the elaborate sequence–of–events necessary to create an unconformity trap, they are surprisingly common and include some of the world's largest fields.

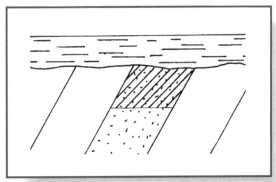

FIG. 2-23   Angular Unconformity

Significance of trap type in the exploration process. Seismic surveying methods are the dominant oil–finding tools used today. Seismic's strength is in revealing subsurface deformation, so the world's structural traps have been rather intensively developed. Stratigraphic traps do not involve deformed rocks, so they not as readily visible to seismic techniques and have not been as intensively developed.

Reservoir Pressure

The fluids in oil and gas reservoirs exist under pressure. When a well is drilled into a reservoir, it creates a conduit to the surface. The pressure differential between reservoir and the surface then drives reservoir fluids horizontally through the reservoir to the wellbore and then vertically up the hole. Reservoir pressure is therefore the principal reservoir drive mechanism or source of energy for producing hydrocarbons.

Reservoir pressure also influences the quantity of hydrocarbons present in the reservoir—particularly in the case of gas because of its compressibility. If a gas reservoir is at twice the pressure of another gas reservoir, it will contain roughly twice as much gas in the same amount of pore space.

Reservoir pressure is also of concern to drillers. If they drill into a reservoir with the weight of their drilling mud less than the reservoir pressure, a blowout can occur.

Where does reservoir pressure come from? When dead plankton and algae sink to the ocean floor, they are under pressure from the weight of the saltwater column above them. Since seawater has a pressure gradient of roughly 1/2 psi (pounds per square inch) per foot of depth, the pressure existing in 3000 ft. of water, for example, is about 1500 psi.

As the organic material is covered by clastic sediments, the particles of sediment stack up on one another in a loose structure—loose enough for sea water to move freely through it. As the sediment thickness builds up—becoming thousands of feet thick—its increasing weight progressively compresses the loose structure, squeezing out entrained seawater. As the weight of the overburden lithifies the clay particles into shale, permeability is no longer measurable, but there still is enough seepage over time to prevent the fluids from taking on overburden pressure.

A useful analogy is large glass marbles in a water–filled goldfish bowl. It's clear in Figure 2–24 that the water does not bear any of the weight of the marbles. Instead, the marbles form a self–supporting structure resting on the bottom of the bowl. The goldfish therefore bears only the weight of the water above him—not the marbles. The marbles are analogous to sediments and the goldfish to reservoir fluids, so fluid pressure in the reservoir continues to reflect only the weight of the fluid column. Reservoir pressure normally encountered around the world is therefore roughly equal to 1/2 psi per foot of depth. This is a very important and useful fact.

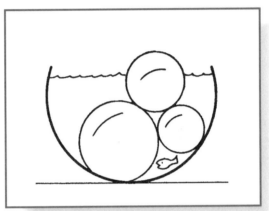

FIG. 2–24 Gold Fish in Bowl with Marbles

Abnormal pressures. Some reservoirs do have abnormally high pressures—sometimes double normal levels—and it's critical that drillers be able to anticipate these situations. The high pressures result from fluid becoming trapped in the rock's pores so that it supports some or all of the weight of the overlying sediments. Several situations can cause abnormally high pressures:

1. Deeply buried, unconsolidated sediments often slump downward toward nearby ocean basins. In moving deeper, the sediments take on additional overburden pressure. This rock pressure is picked up by the entrained fluids because limited permeability prevents the fluids from escaping quickly enough. In time—certainly in geologic time—the pressure will bleed off, but in the immediate time–frame of the driller there is a serious problem.

2. The folding and fracturing associated with major mountain–building (such as the Himalayas) can create over-pressures in mountain–front areas. Again, the pressures will bleed off given enough time.

3. Overpressure can result from artesian flow. This occurs when a formation containing a hydrocarbon reservoir outcrops at a higher elevation where it is charged with water. The reservoir pressure then reflects the weight of a column of fresh water extending upward to the level of the outcrop (FIGURE 2–25).

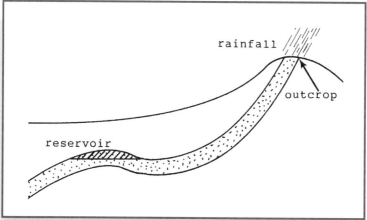

FIG. 2–25   Artesian Overpressure

4. When chalk formations (formed from shells of microscopic marine animals) are compressed during lithification, the individual pores tend to seal off. This traps the fluids which then pick up some of the overburden weight.

5. Sandstones crush to some degree when buried deeper than 15000 ft., which tends to seal off the pores. Very deep drilling therefore typically encounters some overpressure.

Abnormally low pressures are also encountered. This is generally caused by the reservoir having been drilled and produced earlier, resulting in its pressure depletion.

PETROLEUM EXPLORATION 3

Exploration Geology

High-risk, high-cost

The essence of the exploration geologist's job is to put a small "x" on a map that covers a huge piece of the earth's surface. This is the spot where a wildcat well will be drilled to evaluate the petroleum potential of the prospect.

The precise location of the "x" is critically important. Since the hole to be drilled is only about eight inches in diameter, it's easy to miss the target, and a single unsuccessful wildcat (dry hole) often condemns the entire prospect.

Although precision is essential, it isn't really possible because exploration geology operates with a minimum of information. Only in the later development phase does a lot of information become available.

As a result, most wildcat wells are "dry" (not producing commercial quantities of oil or gas), with the cost of the well—millions of dollars— entirely lost. A more serious loss occurs when incorrect placement of the wildcat caused a potential new field discovery to be missed. After the prospect is abandoned by the original operator, it is not unusual for a competitor to pick it up and drill the discovery well.

Exploration geology is clearly a high–risk, high–cost enterprise—not a place for the faint–hearted.

A process of elimination

Exploration geologists start out with the entire world as a potential target. They then go through a process of elimination to reduce the area of interest to manageable proportions.

Much of the world's surface has no petroleum potential because igneous or metamorphic basement rocks are exposed at the surface. Exploration is therefore restricted to the world's sedimentary basins where sedimentary rocks lie on top of the basement rocks (FIGURE 3–1).

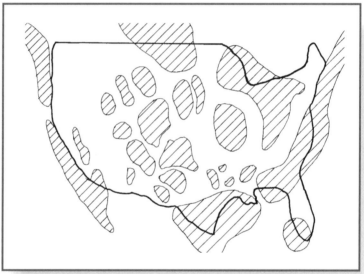

FIG. 3–1   Sedimentary Basins in the United States

Selecting the specific basin on which to focus exploration is called regional geology. Once a basin has been selected, the more focused process of picking a well site is prospect geology.

When evaluating the potential of a virgin, undrilled basin, one concern of the geologists is whether adequate volumes of black marine shale or other source rocks are available. Once commercial oil is discovered in a particular basin, however, future exploration can proceed without concern for source rocks.

Exploration in the past

Oil seeps. There are many oil seeps around the world where black, tar–like oil oozes from sedimentary rocks exposed at the surface. Indigenous peoples have always gathered this tar for fuel and other uses.

In the early days of the oil industry, very little was understood about geology, but wildcatters found that drilling near seeps sometimes discovered reservoirs of oil. This oil was often much lighter and more volatile than the tars found at the surface. It therefore yielded more of the kerosene–range lighting products most valued at that time.

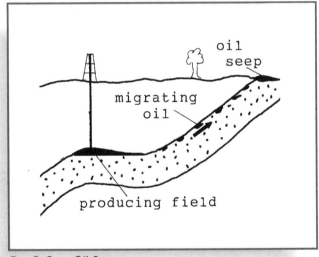

FIG. 3–2   Oil Seep

Geology provides an understanding of oil seeps and associated commercial reservoirs. Figure 3–2 illustrates oil droplets migrating upward through an inclined porous and permeable formation. If the migrating oil encounters a structure with closure such as the anticline shown, it first fills up the structure, then spills over and continues its upward migration until it reaches the surface where it creates an oil seep. The lighter, gaseous components in the exposed oil rapidly dissipate into the atmosphere, leaving the non–volatile, tarry residue typical of seeps.

Surface geology

In the early days of the industry, petroleum geologists spent much of their time in the field doing surface geology. They were a hardy group, traveling and camping in remote deserts and jungles for months at a time.

Surface geology focuses on formation outcrops—surface exposures of the underlying rocks. When an outcrop is located, its location is surveyed so that the data from it can be used to map the area. The rock is examined to determine lithology and samples are taken for later laboratory examination. If this is the first survey of the area, particular note is made of potential reservoir and source rocks.

The dip, which is the angle of inclination of the rock's bedding planes, is measured with a Brunton compass, as is the direction of the dip (azimuth). Structural maps are then constructed by plotting the dips and strikes (horizontal lines perpendicular to the dips) on the map. Anticlines and synclines in the area can then be identified, with the anticlines being candidates for exploratory drilling (FIGURE 3–3).

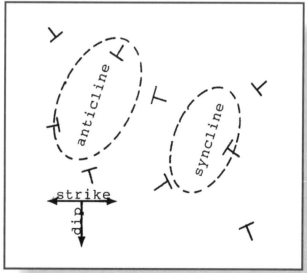

FIG. 3–3   Surface Geology Structural Map

Surface geology plays only a minor role in modern exploration. The large, obvious structures that it could find have largely been drilled. The search for the more subtle traps that remain requires more sophisticated approaches—particularly modern seismic techniques. Today's exploration geologist is much more likely to spend his time sitting in front of a computer than slogging through swamps.

Early geophysics

Geophysics—the physics of the earth—contributed some useful tools for oil finding at an early stage of the industry. These were the "gravimeter" and the "magnetometer"—extremely sensitive devices to measure the earth's gravitational pull and magnetic field. These instruments are used to detect deep–seated igneous intrusions (basement highs) that can be the foundation for anticlinal petroleum traps when sedimentary formations drape over them.

Igneous material is typically denser than sedimentary material, so the closer the intrusion comes to the surface, the higher the reading on the gravimeter. By taking scattered readings in an area, and plotting the readings on a map, high–gravity anomalies that indicate the top of the structure can be identified. The gravimeter can also locate salt domes which appear as low–gravity anomalies.

Magnetometer surveys are conducted similar to gravimeter surveys. They detect anomalous distortions of the earth's magnetic field by the presence of iron–rich igneous material. The closer the intrusion is to the surface, the stronger the anomaly.

The efficiency of both tools was greatly increased when airborne sondes were developed. This permitted quick and inexpensive surveys of broad areas. These tools are still used to some extent today, but they have been largely displaced by the more definitive (and more expensive) seismic techniques.

Seismic exploration

Seismology is the study of earthquakes. It involves the measurement and interpretation of acoustic waves radiating outward through the earth from earthquake sites. To apply seismology to the petroleum industry, explosives or other means are used to create miniature earthquakes that help locate structural traps—those formed by deformation of the rock formations. Anticlinal traps and fault traps are examples.

Basic principles.   In Figure 3–4, the seismic crew generates an acoustic impulse at the surface by an explosion or other means. This sends an acoustic pulse radiating downward through the rock formations. When the pulse encounters a sharp change in acoustic characteristics, called a marker bed, a portion of its energy is reflected back to the surface. Note that the angle of incidence on the marker bed is identical to the angle of reflection. The remaining acoustic energy continues downward.

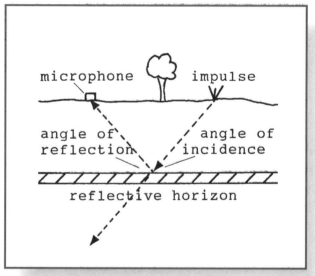

FIG. 3–4   Seismic Geometry

The return of the reflected energy to the surface is recorded by a microphone, and its transit time down to the reflecting horizon and back to the surface is calculated. The locations of the impulse and the microphone have been accurately surveyed, so the subsurface distance traveled by the energy pulse can be calculated using an average acoustic velocity for the rocks. The depth and location of the reflection point is then calculated by assuming that the reflection point is located midway between the shot point and the microphone.

Figure 3–4 illustrates one ray of the acoustic energy which radiates downward at all angles from the explosion. Additional rays strike other locations on the marker bed (FIGURE 3–5) and reflect back to different microphones. In this way the structure of the marker bed is defined.

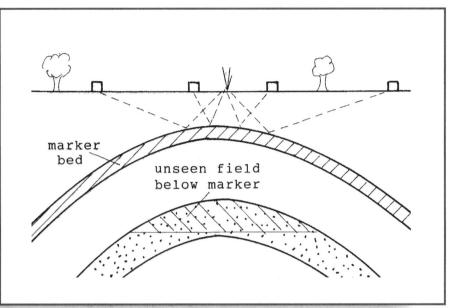

FIG. 3–5 Multiple Reflections Define Structure

Marker beds are not reservoir rocks; they are simply formations whose structure can be clearly defined by seismic methods. Reservoir rocks in their vicinity, however, can be expected to conform to the same structure. If an interesting structure is noted in a marker bed, a wildcat well may be drilled to test the unseen porous and permeable formation known to be located a few hundred feet above or below the marker.

3–dimensional seismic surveys. In recent years, the use of 3–dimensional seismic surveys has increased rapidly and now dominates the activity. 3–D seismic differs from the earlier 2–D seismic in two essential respects:

- 3–D shoots a much tighter grid, greatly increasing the definition of the prospect.

- The data is used to construct a 3–dimensional mathematical model of the prospect on a mini–computer. The analyst can then make electronic cuts across the prospect at any angle.

3–D surveys are costly, but have proven very effective in defining complex structures and reservoirs in both exploration and development operations. Its additional costs are much more than offset by the additional oil found through better understanding reservoir geometry, by drilling fewer dry holes and by optimal placement of offshore platforms.

4–dimensional seismic surveys. A relatively new concept, 4–D involves time as the fourth dimension. A reservoir is re–shot on a regular basis to monitor the progress of enhanced recovery projects or natural drives.

OBC cable. On–bottom–cable (OBC) seismic utilizes static hydrophones attached to cables laid on the seafloor. The technique is applied to densely developed fields where the many platforms interfere with normal streamer–cable operations. It is particularly useful in 4–D work.

Seismic Phase I: Data Gathering

The first phase of seismic exploration is the running of seismic surveys to gather basic data. Oil operators usually contract this service out to specialized seismic companies that are equipped to work in any environment, anywhere in the world. When a specific tract is of general interest to the industry, however, the contractors often do speculative shoots at their own expense and sell the data to multiple operators.

Land operations. The acoustic impulse has traditionally been generated with dynamite or other explosive detonated in a shot hole. However, in countryside where off–road vehicles can operate, most work is now done with truck–mounted vibroseis units. The truck pulls up to the shot point, drops a steel pad from its underbelly, jacks itself up on the pad until its rear wheels are off the ground, then vibrates the pad to generate shock waves. Vibroseis is faster and cheaper than explosives and has environmental advantages.

Shooting the seismic line is a nonstop process, laying down microphones, called "jugs," for the upcoming shot, while picking up jugs from the previous shot. A group of several jugs is used at each recording point. The locations of the recording points and the shot points must be precisely surveyed, which may require the clearing of trees and brush. Costs are greatly increased when natural obstacles such as swamps, ridges, rivers, and thick forests are crossed.

Three–dimensional data is gathered by stringing multiple microphone lines on either side of the shot line.

Offshore operations. Marine operations are much faster and cheaper than land operations (FIGURE 3–6).

Groups of hydrophones (underwater microphones), are embedded in long streamer cables towed behind the seismic vessel at considerable speed. Either explosives or a sudden release of compressed air from an airgun is used to generate the acoustic pulse.

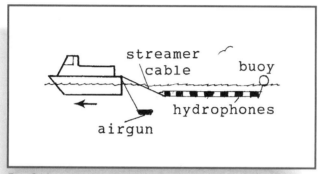

FIG. 3–6   Marine Seismic Survey

The ship's computerized navigation system is locked onto a satellite and calculates the precise position of the shot points and hydrophones continuously. The on–board computers map the structures in real time as they are being traversed. If an interesting feature is spotted, the ship can circle back and run cross–lines to get more detail.

Marine surveys usually yield seismic records of excellent quality because water is the ideal medium to transmit shock waves.

Three–dimensional data is gathered by running two or three ships in tandem, each pulling multiple hydrophone streamers.

Seismic Phase II: Processing

The second phase of seismic exploration is processing the raw data to refine it into a useful form. Processing is done by geophysicists, and they use very large computers—often supercomputers—to manipulate the huge amounts of data. Seismic is by far the most demanding computer application in major oil companies.

Seismic processing is extremely critical to an oil company. If done well, it may discover a huge new field and revitalize the company. If done poorly, the field may be missed. Throughout the history of processing, various companies have at times developed new processing techniques that allow them to locate new fields that are invisible to the industry in general. This gives them an advantage for a time, but the industry catches up quickly.

Because it is so critical, large oil companies tend to process most of their data internally and use many proprietary routines. Smaller companies, however, often contract out to commercial processors.

Some of the normal processing routines are:

- *Static correction,* which adjusts for the near–surface effects of loose soil, sand dunes, permafrost, etc.

- *Migration,* which corrects the data for tilted formations. The basic algorithm assumes the two legs of a reflection are of equal length; but as can be seen in Figure 3–5, this is not true for tilted formations.

- *Bright spot processing,* which is a direct detection technique (DDT). This means that it can detect the presence of sub–surface hydrocarbons before a well is drilled. Bright spots are highly visible anomalies on seismic profiles (cross–sections) that indicate the presence of gas–filled porous rocks below. The anomaly is caused by the acoustic velocity being slower in the gas than it would be if the porosity were either filled with liquid or if there were no porosity.

  In provinces like the U.S. Gulf Coast and the North Sea that typically have gas–caps associated with the oil reservoirs, the gas–caps may be detectable by bright spot techniques. Drilling for oil can then proceed with confidence that an oil leg will be found below the gas cap.

  Another use of bright spots is to make certain that off-shore drilling locations are free of shallow gas zones that could blow out and catch fire when the well is spudded. This is a major problem, for example, in the southern part of the South China Sea.

- *Dim spot processing,* another DDT. The anomaly on the seismic profile is caused by hydrocarbon gases leaking from incompletely sealed traps. The gases percolate upward, displacing water in the porosity, and slow the acoustic velocity in the overlying formations. On the seismic profile this causes a distinct "chimney" or "halo" effect above the trap.

Seismic Phase III: Interpretation

The third and last phase of the seismic process is interpreting the processed seismic profiles to determine whether a well should be drilled. Interpretation is usually performed by geologists, although geophysicists sometimes do it. The interpreter studies the seismic profiles and translates them into anticlines, synclines, faults, and reservoirs seen by bright spots, etc.

Other geologic tools

In oil and gas provinces where the traps are structural and where seismic works well, picking the spot for a wildcat is often quite simple—particularly if some DDT is available. For stratigraphic plays, however, there is no easy solution. In this situation, geologists use a variety of less direct tools to try and understand the subsurface. Some of these are discussed below.

Following trends. If, for example, two fields have already been discovered, the geologist may connect them with a trend line and then extend the line to spot the next wildcat. It must be kept in mind, however, that curved lines join two points as well as straight lines do. There can therefore be little confidence placed in any single trend line that is picked (FIGURE 3–7).

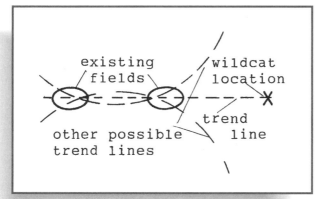

FIG. 3–7   Spotting Wildcat by Extending Trend

Projecting backward in time. Geologists project themselves into ancient time, visualizing the mountains and rivers and seas that existed when the target formation was being deposited. If, for example, the first wildcat that was drilled on a prospect encountered shale in the target interval, the next wildcat would be moved towards the ancient high. This would improve the chances of its encountering sands that had been deposited close to the ancient shoreline (FIGURE 3–8).

The sequence of ancient events—the order in which they occurred—can be the key to whether a prospect will be productive. For example, if a structural trap identified by seismic was not yet formed when the oil or gas migrated past that point, the trap would be empty. Geologists therefore are very concerned with dating both migration and trap formation.

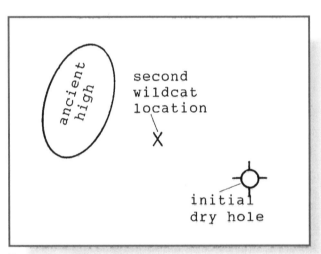

FIG. 3–8   Wildcat Strategy

Geologic mapping.   Geology is a spatial science, concerned with which rocks occupy what spaces, so mapping is fundamental to its practice. As pictures created from the numeric data, maps help the visualization of subsurface structures. They also provide an organized framework for interpolating between and extrapolating from the widely scattered data points. As a result, maps are powerful interpretive tools.

Maps are displayed either on paper or on computer screens, both of which are in two dimensions. The geologist, however, must be able to visualize prospects in three dimensions, so views at right angles to one another are used.

Plan views are from the perspective of an observer looking vertically downward. Outcrop and contour maps are typical plan–views.

Cross–sectional views (profiles) are from the perspective of an observer looking horizontally. They can be constructed either from seismic or from well data. The most frequently drawn maps are contour maps using seismic or well data to define the structure of subsurface horizons.

The depth data is plotted on the contour map with reference to sea level. For example, the depth of a 15,000 ft. well drilled from a surface location 5,000 ft. above sea level is plotted as –10,000 ft.; that is, as 10,000 ft. below sea level. A 22,000 ft. well drilled from an elevation of 12,000 ft. above sea level is also plotted as –10,000 ft. Wells that bottom out above sea level are plotted as positive numbers (FIGURE 3–9).

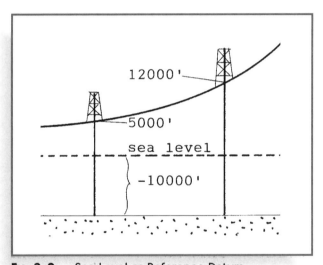

FIG. 3–9   Sea-Level as Reference Datum

By using sea level as the common reference, the effect of surface topography is eliminated, and the depth of subsurface horizons in different wells (or shot points) are related directly to one another.

When all the data points are plotted, contour lines are drawn using an appropriate interval, say every 100 ft. Where a given contour line passes between two data points, its relative distance from each point is interpolated. In Figure 3–10, the –6400 ft contour passes between the –6580 ft point and the –6340 ft point. Since the contour is 180 ft above –6580 ft and only 60 ft

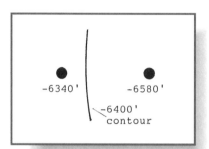

FIG. 3–10   Interpolation Between Data Points

below –6340 ft, it is drawn much closer to the –6340 ft point (25% of the total distance between the two points) than to the –6580 ft point.

When the contouring is completed, the structural highs stand out as 3–dimensional mountains and the lows look like valleys. The closer the contour lines are squeezed together, the steeper the slope (FIGURE 3–11).

The rules of proper contouring are:

- Contour lines never cross

- Straight lines are rare in nature, so most contours are smoothly curved

- Each contour line should reflect the character of the contour lines above and below it

Proper contouring is very demanding, but the effort is worthwhile because it can reveal prospects that would otherwise be overlooked. Computer–generated maps also need to be adjusted by hand to assure that the interpretive possibilities are fully realized.

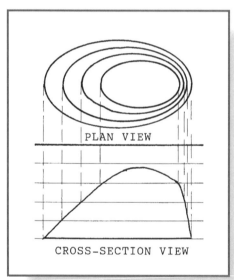

FIG. 3–11   Contour Mapping

CONTRACTS AND REGULATIONS 4

Land Ownership

In the United States, land ownership has two separable portions—surface and minerals. Either the surface rights or the mineral rights on a particular tract can be bought, sold, or leased separately or together. When the surface and mineral rights are owned together, it is called ownership in fee.

Most land in the United States is owned by private individuals or companies, but state and federal governments also own considerable land. In some western states and Alaska, the federal government owns most of the land.

Offshore, most states own the mineral rights from the shoreline to three miles out. Texas and Florida own out to 3 leagues, or 10.5 miles. The federal government then owns from the state boundary to the edge of the continental shelf (defined as 200 miles out or until a water depth of 8500 ft is reached). These federal lands are referred to as the outer continental shelf (OCS).

Only the United States and portions of Canada have private ownership of minerals. Throughout the rest of the world, the state owns all minerals.

Acquiring Mineral Rights in the United States

Mineral leases

Although the mineral rights under a parcel of land can be purchased outright, oil companies usually obtain the rights to explore for and develop oil and gas by executing an oil and gas lease with the landowner.

The legal description of the leasehold specifies both its surface area and its vertical dimension. Most oil and gas leases are unlimited vertically, but some are restricted to specific pay zones or intervals, i.e., 4000 ft to 8500 ft.

The terms and conditions of government leases are usually fixed—not subject to negotiation. They are often awarded by competitive bidding, particularly in areas where industry interest is high. The huge lease sales conducted by the Minerals Management Service for tracts on the OCS, for example, yield hundreds of millions of dollars in competitively bid lease bonuses. Government lands of lesser interest, such as most onshore leases, are usually awarded non–competitively on a first–come, first–served basis.

When dealing with private landowners, typically farmers or ranchers, oil companies negotiate each provision of the lease. If there is little drilling activity in the area, the lease will be cheaper than if the area is active. If an active lease play is underway involving a number of companies, the landowner is in a strong bargaining position and can extract costly concessions from the company.

Landowner negotiations are conducted by landmen, who are often lawyers or paralegals. They are also selected for their inter–personal skills that facilitate reaching agreement with the landowner at a reasonable price. Some of the major provisions in lease agreements are:

- *Primary term.* The mineral rights are normally leased for a fixed duration, say five years. If commercial production is not established by the end of this primary term, the lease expires. If commercial production is established, the term is automatically extended—held by production—until abandonment of the field.

- *Signature bonus.* The landowner usually receives a signature bonus upon signing the lease agreement. The size of this bonus varies widely depending on the intensity of leasing activity in the area and how effectively each party negotiates. For example, land far from any oil activity can be leased for bonuses as low as a dollar an acre. On the other hand, bonuses of thousands of dollars an acre are regularly paid in more active areas.

- *Royalty.* The landowner's most significant financial return from the lease is the royalty. This is the fraction of the revenue from each barrel of oil or cubic feet of gas sold that is to the landowner's account. Since royalty is a carried interest, the landowner bears no development costs or operating expenses.

  Historically, the landowner's royalty has been one–eighth, and this is still the most common fraction. In recent years, however, some landowners have been able to negotiate higher royalties in hot leasing plays. One–fourth, or even one–half royalties appear at times. On the OCS, the federal government usually retains one–sixth.

- *Drilling obligations.* The oil company may commit to drill a specified number of exploratory wells to a specified depth each year during the primary term. If the wells are not drilled, the lease terminates. Some leases have a third option wherein the lease can be held by delay rental payments paid to the landowner in lieu of drilling the wells.

- *Surface access.* The lease provides the oil company with the right of reasonable surface access to conduct operations. This includes building roads, drilling locations, tank battery pads, etc. The landowner receives surface damage payments to compensate for the revenue lost while the affected land is unavailable for agriculture or other use.

Later transactions. Once a lease is obtained, the oil company has several options in exploiting it:

- *Develop the property alone.* If the company geologists are particularly enthusiastic about the prospects of finding oil or gas on the property and adequate funds are available, the company may proceed to drill the wildcat wells necessary to test it and, if successful, to develop it.

When additional capital is needed, operators often obtain loans from banks or other sources with production payments—pledging a fixed percentage of the revenue stream from other producing properties to repay the loan.

- *Take on a partner.* If the capital requirements are uncomfortably large, the company may seek a partner or partners to share the costs in proportion to their revenue interests. One partner, usually the one with the largest stake, functions as the operator and performs the actual work. The non–operating partners pay a fee for these services.

- *Seek a farm–out.* The company may have other drilling prospects that it rates higher in terms of the potential profit versus the risk. It therefore may choose to spend its available capital on the other prospects and farm out this one to another operator.

  Farming out is essentially a risk–reducing financing mechanism. When Operator "A" farms the property out to Operator "B", "B" bears all drilling costs while "A" retains some form of interest in the revenue stream if oil or gas is found. This might be an over–riding royalty interest—say one–eighth of "B"s seven–eighths. An over–ride, like landowner's royalty, is a carried interest. Other options are for "A" to retain a net profits interest or to back–in to a working interest of, say 50%, if a discovery is made.

  Another farm–out mechanism is for "A" to divide the lease into a checkerboard pattern and allow "B" to earn full rights to every other square by drilling a well on it. This shifts the exploration risk to "B". If a discovery is made, however, it's highly likely that the field will extend onto the offsetting squares retained by "A". "A" can then proceed directly to development drilling without having to drill the more risky exploration wells (FIGURE 4–1).

- *Drilling funds.* Some oil companies form drilling funds by recruiting a number of investors as limited partners. The

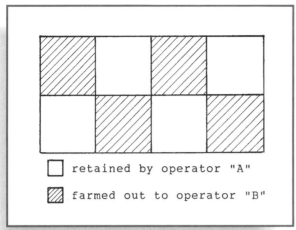

FIG. 4–1   Checkerboard Farm-out

investors earn a revenue interest but have no voice in operating decisions. Their liability is also limited.

Drilling funds are not nearly as popular since the incremental income tax percentages and percentage depletion have been reduced.

In their heyday, serious excesses occurred as promoters preyed on naive investors. When the general partner sells his revenue interest in the well(s) to limited partners, he is no longer concerned with the well(s) long term economic viability. Instead, he will view wells primarily as promotional tools. This has lead, for instance, to wells being drilled near the road to attract potential investors. One aim is to drill where enough oil or gas will be produced initially to encourage investors; whether enough is produced over the well's life to be profitable is far less important.

It is fair to say that drilling funds are fundamentally flawed investment vehicles because of the essential differences of interest between the general and the limited partners.

Regulatory Oversight

The development of an oilfield overseas typically involves only a multinational oil company and the host country's national oil company. The regulatory function is quite straightforward, consisting largely of contract enforcement.

Developing a similar field onshore in the United States is complicated by multiple operating companies and dozens of private landowners. The regulatory requirements are correspondingly more complex.

Each of the producing states has an oil and gas regulatory agency. In Texas, it is the Texas Railroad Commission; in Oklahoma, it is the Oklahoma Corporation Commission; etc. The Minerals Management Service regulates activity on federal lands.

The various functions of regulatory agencies are described next.

Protecting landowner's correlative right

The Rule of Capture. When a well is drilled and produced, an area of lower pressure is created in the immediate vicinity of the wellbore. This establishes a pressure differential between the rest of the reservoir and the wellbore, which causes migration of reservoir fluids towards the wellbore.

Since petroleum could migrate across property lines, the rule of capture was applied. This is a legal concept unique to oil and gas and the taking of game animals. The law conveys title of oil and gas to whoever drills a well and produces it, even if the produced oil and gas may have migrated from an adjacent property.

Slant holes. In the frenzy of the East Texas Field drilling boom, unscrupulous operators drilled slanted holes from surface locations beyond the field's limits to tap into the reservoir beneath their neighbor's property. Regulation was therefore imposed to prevent this hot oil production by requiring a survey of the bottom–hole location of each well drilled.

The Offset Drilling Rule. If a reservoir extends across a property line, one landowner could drill a well close to the line and suck oil and gas out from under his neighbor's property. The neighbor's only recourse is to immediately drill an offset well on his side of the line to equalize the pressure, preventing migration.

When land is leased, the offset drilling rule now obligates the lessee to immediately drill an offset to any producing lease line well. When there is a question whether the reservoir actually extends across the lease line, the lessee might otherwise have chosen not to drill the well until more information was available. The offset rule therefore causes some dry holes, but does protect the rights of the lessor.

Maximizing efficient recovery of hydrocarbons

A problem with the rule of capture is that it provided an incentive to produce as much as possible before it was captured by others. This encouraged a frenzy of drilling; sinking wells as quickly as possible and often right beside one other. The produced gas was flared to the atmosphere and wasted, and the rapid blow–off of reservoir energy reduced the ultimate oil recovery.

A new level of regulation was therefore initiated under the banner of maximizing the efficient recovery of hydrocarbons. The rules were designed both to improve recovery and to eliminate the wasteful drilling of more wells than was necessary to deplete the reservoir. The regulatory agencies thereby assumed responsibility for optimizing profitability of oilfield development as well as maximizing hydrocarbon recovery.

Well spacing rules. A minimum number of acres per well was specified. For example, 40–acre spacing, a typical spacing for oil wells, means that one well can be drilled on each 40–acre drilling unit. Although somewhat arbitrary, the assignment of 40 acres is based on a rough reservoir engineering determination that 40 acres is the maximum one well could adequately drain. This can be better understood in the typical assignment of 640 acres (one square mile) as the spacing for gas wells. Since gas has minimal viscosity as compared to oil, one well can drain a much larger area.

Interestingly, the deeper the producing horizon, the wider the spacing. This does not reflect a reservoir reality, but rather an economic reality since deeper wells are more costly. Again, the dual aims of regulation are clear.

Applications for a drilling permit must include a plat (map) showing the dimensions of the drilling unit committed to that well, plus any other drilling units nearby that are already approved. Drilling units are normally square, with the well drilled in the center, but some deviation is permitted in the shape of the unit and the well location.

Where the land is broken up into relatively small individual parcels, it is often necessary to combine two or more parcels to assemble a drilling unit that is large enough. This is particularly common when assembling 640–acre gas well units. This is called pooling, or unitization, and the operators involved share costs and revenues in proportion to the share of the unit they contributed.

Minimum distance from property line. The regulatory agency specifies how close wells can be drilled to property lines; 660 feet is typical. The purpose is to minimize migration across lease lines.

Well allowables. The regulatory agency sets well allowables for each field. A well's allowable is the maximum barrels per day it is permitted to produce.

Allowables are designed to maximize recovery by equalizing the pressure drop throughout the reservoir and preventing the reservoir from being depleted too fast. Economics are also considered, with deeper wells receiving higher allowables.

High gas–oil–ratio wells are given lower allowables. This is because the gas produced also depletes reservoir pressure and allowables are designed to equalize the reservoir depletion caused by each well.

Some new wells produce less than the lease's maximum allowable, so they are assigned an allowable equal to what they produced on a 24–hour well test. New wells that initially received top allowable are assigned progressively lower allowables as the reservoir pressure decline reduces the well's productivity. This is determined by monthly well tests. In a given month, a tank battery can produce the sum of its individual well allowables.

Because of the maturity of oilfield development in the United States, very few wells are capable of producing more than their top allowable. This says that the United States has essentially no surplus producing capacity.

Proration. The Texas Railroad Commission stabilized world crude prices for several decades through proration of Texas' oil production. The mechanism used to limit production was to prorate the number of producing days in a month. That is, the month of August might be assigned ten producing days.

This says that in August a total of 10 days' allowable could be sold from each tank battery.

Texas has not been in a position to prorate production for some time. For the last twenty years, Texas and the entire U.S. industry has essentially been producing at capacity with the country a net importer of crude. At the moment, the United States imports more than half of the crude oil it consumes.

Other regulatory functions. Before a new secondary or tertiary recovery project is initiated, the regulatory agency conducts a hearing to determine whether the action is appropriate with respect to reservoir mechanics and economics, and to assure that correlative rights are protected.

A major project usually involves unitizing multiple land holdings. A single operating unit is formed with each landowner receiving equity in proportion to the value of its contribution. The numerous reservoir engineering assumptions involved in the assessment of relative value are reviewed by the regulatory agency.

Regulatory agencies perform a classification function in defining field limits both laterally and vertically. Vertical definition is particularly demanding because of the variability of reservoir rocks. For example, a somewhat arbitrary decision often must be made whether to include an isolated porous zone as part of the main pay, or whether it should be considered as a separate reservoir. In the first case, the zone would be commingled with the rest of the pay at little additional cost. In the second case, the operator bears the additional cost of a separate completion.

Avoiding hazards to health & environment

- Specifying casing and cementing standards plus abandonment procedures to protect freshwater zones.

- Setting environmental protection standards regarding gas flaring; disposal of drill cuttings; oil & saltwater spill containment and clean–up, etc.

- Setting safety standards regarding toxic hydrogen sulfide gas, chemicals handling, safety control systems, etc.

International Petroleum Contracts

Host country objectives

The host country needs the operating company's expertise and financial capital to unlock their hydrocarbon resources. They are also interested in jobs that petroleum operations provide for nationals.

Expertise. The oil company's expertise is needed to provide the petroleum technology and the management of petroleum operations. Equally as valuable is the entrepreneurial approach used by the company is to succeed in this high risk business.

Financial capital. The host government has difficulty funding petroleum development and looks outside for capital because

- The industry is extremely capital intensive, requiring huge sums
- Much of the capital (i.e., exploration) is subject to a high risk of complete loss
- It normally takes several years to find and develop a new oil or gas field, so the capital is invested for a long time without any return

Jobs—nationals vs. expatriates. Jobs often become a source of tension between the host country and the foreign operating company. This is an area where the two parties' interests clearly diverge.

The host country seeks to maximize the number of jobs held by nationals. This is more than an issue of injecting the maximum dollars into the local economy; it is an issue of national pride. They are particularly interested in placing nationals in the higher ranking managerial and professional jobs.

The companies, however, have a strong interest in keeping expatriates in the key jobs. Trusted "expats" are deeply experienced in the unique problems of operating in remote places. With their years of exposure to the company's way of conducting business and their wide acquaintance with company personnel throughout the world, they are uniquely able to get the job done. It's difficult for any national, however talented, to match these attributes.

In the long run, the solution is to recruit able nationals for the company's mobile expatriate corps. That way, it's not necessarily Brits or Americans that are brought in for the key jobs. This defuses much of the resentment.

In the short run, however, the problem will continue with the host country trying to set restrictive quotas on the number of expatriates.

Operating company objectives

The operating company's interest is in establishing a profitable new business enterprise. The critical elements are

- *Geologic potential.* The contract area must offer an attractive probability of finding oil or gas in commercial quantities.

- *Fiscal terms.* The agreement must provide terms generous enough to justify the risks involved. It's essential that the company can freely take cash out of the country.

- *Political risk.* The company must be confident that the terms of the agreement will be honored over time by the host country.

Nationalization by the host country

A critical moment is when production and its associated revenues first begin. The company has already invested its capital and provided the expertise to find and develop the field. This can tempt the host government to expropriate the company's holdings at this point.

Host governments have a number of expropriative tools other than outright nationalization: the tax rate can be raised, the reference price for crude oil to which the tax rate is applied can be raised, production can be restrained, or tariffs & license fees can be imposed. Any of these more subtle measures can quickly erode the company's profit to assure that it leaves.

Operating companies must never forget that host governments come under extreme pressure to eject any foreign company that is perceived to be profiting excessively—taking out much more than they are putting in. It doesn't matter whether the country is a democracy or a dictatorship, the pressure will be there and the political leaders are vulnerable to it.

The most secure situation is where there is an ongoing need for the company's expertise; for example, where development is moving into deeper and deeper waters, requiring ever–new technology. Even in this situation, however, the company must accept that their profits will be modest. In fact, they must take action to assure that they are modest if expropriation is to be avoided.

Types of agreements

The agreements now in effect between host countries and operating companies differ in each case, but fall within the following three general categories:

- *Concession contracts.* The operator makes the investment decisions; manages the day–to–day operation; pays all exploration, development and operating costs; and sells the produced hydrocarbons.

  The government has a royalty interest, receiving payment from the operator for the revenues generated by the sale of a specified percent of the produced oil and gas. Royalty is a carried interest, bearing none of the costs. The government also taxes the operator's profits.

  Concession agreements are used in the United States, Canada, and the North Sea.

- *Production–sharing agreements.* As with the concession contract, the operator manages the operation, pays all costs and sells the production.

  The operator's investment is then recovered from the sale of cost oil, a pre–agreed portion of the production. Receipts from the sale of the remaining production—profit oil— are split between the operator and government according to the terms of the agreement. The government also taxes the operator's profits.

  Some agreements give the operator higher portions of the profit oil for small fields than for large fields. This has the dual effect of encouraging the development of small fields while reducing the operator's profits from very large fields. This latter can actually benefit the operator by preventing resentment from developing over excessive profits—which can lead to expropriation.

- *Service contracts.* The operating contractor is paid on a $/bbl basis to conduct the operations under the close supervision of the government—usually through the national oil company. The government provides all capital and the contractor has no ownership interest in the oil and gas produced.

RESERVOIR PERFORMANCE 5

Reservoir Fluids

Definitions

Fluids. Any substance that flows and yields to any force tending to change its shape is a fluid. Both liquids and gases are fluids.

API gravity. API gravity is the density measure used for petroleum liquids. It is expressed in degrees. The higher the API gravity, the lighter the liquid. The relationship to specific gravity (S.G.) is

$$API\ gravity = 141.5/S.G. - 131.5$$

Petroleum gas. A petroleum compound is defined as a gas if it is in a gaseous state at normal atmospheric conditions (60° F and 14.7 psi in the United States). For example, butane is considered a gas while pentane is considered a liquid.

Solution gas. Solution gas is dissolved in the oil under the initial reservoir pressure and temperature. Under these conditions, the gas is in the liquid state, not the gaseous state.

Critical saturations. The minimum saturations of oil, water, or gas in the reservoir that cause the fluid to be a continuous medium, and therefore producible (see "Relative permeability", p. 29).

Bubble point pressure. When a reservoir is above its bubble-point pressure, it has no free gas—all gas is in solution in the oil. As the reservoir is produced and pressure declines, the bubble-point pressure is reached. Gas comes out of solution, forming a free gas saturation.

Condensate. Condensate is light hydrocarbon liquid formed by condensation of petroleum compounds that were in the gaseous phase under initial reservoir conditions. It is highly volatile and from clear to a light yellow in color.

Viscosity (μ). Viscosity is a fluid's resistance to flow. The thicker & stickier the fluid, the higher its viscosity. Oil viscosity is measured in centipoise (cp).

Formation volume factor (ß). Beta is the factor for the volume change undergone by the reservoir fluids when they are produced. In the case of oil, it is the ratio of the space occupied by a barrel of oil at reservoir conditions to the space occupied by a stock tank barrel (STB) of the oil. Most oils shrink when their solution gas dissipates at the surface and because of the cooler surface conditions, so their Betas are greater than 1.

Gas-oil ratio (GOR). The producing GOR (R) is the standard cubic feet of gas produced per stock tank barrel of oil produced (SCF/STB).

The solution GOR (R_s) is the standard cubic feet of gas in solution in a stock tank barrel of oil at initial reservoir conditions. R_s may or may not be the same as R.

Gas cap. A gas cap is free gas trapped in the top of the structure above the oil leg. When there is a gas cap, the reservoir is at bubble-point pressure.

Associated and non-associated gas. Associated gas includes solution gas and gascap gas. They both occur only in association with oil.

Non-associated gas is from solely gas reservoirs where no "black" (non-condensate) oil is present.

Recovery factor. The recovery factor is the percent of the reservoir's original oil in place (OOIP) or original gas in place (OGIP) that will be recovered.

Recovery Nomenclature

Primary recovery. Producing the reservoir using only the natural reservoir energy.

Secondary recovery. A process that adds energy to the reservoir following primary recovery. An example is waterflooding a depleted field.

Tertiary recovery. Follows secondary recovery. For example, CO_2 flooding could follow waterflooding, which had followed primary recovery.

Pressure maintenance. The initiation of additional recovery processes like waterflooding immediately without waiting for primary depletion.

Enhanced recovery / improved recovery. Used interchangeably to refer to all the artificial drive mechanisms such as waterflood, CO_2 flood, steam injection, etc.

Fluid systems

Since each reservoir contains a unique blend of many different hydrocarbon compounds, no two reservoirs behave identically when produced. Instead, they cover a wide spectrum of variation.

The following classification of hydrocarbon systems is an attempt to reduce this complexity to workable levels. It is based on the reservoir's relative mix of lighter and heavier hydrocarbons and starts with the heaviest mix.

BITUMEN
$4 < °API < 10$
$1,000,000 > \mu > 5,000$
$\beta \approx 1$
$R_s = 0$

Because it has few small molecules, bitumen's viscosity is so high that for practical purposes it is immovable in a reservoir. Only if the reservoir is shallow enough for a mining/retorting operation can bitumen be commercial. It is therefore only of interest to the mining industry, not to the oil and gas industry.

TAR OR HEAVY OIL

$10 < °API < 20$
$5,000 > \mu > 100 CP$
$1.0 < \beta < 1.1$
NEGLIGIBLE $< R_s < 50$

With more lighter ends than bitumen and a small amount of gas in solution, heavy oil flows sluggishly through a reservoir. To attain commercial production rates, however, it's usually necessary to reduce the crude's viscosity by injecting steam.

LOW-SHRINKAGE OILS

("BLACK OILS")

$20 < °API < LOW\ 30's$
$100 > \mu > 2\ TO\ 3\ CP$
$1.1 < \beta < 1.5$
$50 < R_s < 500$

This oil's viscosity is low enough for it to move through the formation readily, so low-shrinkage reservoirs are normally commercially producible. When production starts, however, the limited quantity of gas in solution blows off quickly and causes reservoir pressure to decline rapidly. As a result, the primary recovery factor is generally low and secondary measures are indicated.

HIGH-SHRINKAGE OILS

("VOLATILE OIL")

$LOW\ 30's < °API < LOW\ 50's$
$2\ TO\ 3 > \mu > 0.25\ CP$
$1.5 < \beta < 2.5\ TO\ 3.5$
$500 < R_s < 2000\ TO\ 6000$

Because of its low viscosity, volatile oil produces at high rates and its high R_s results in good recovery factors. It also contains a maximum of higher-valued gasoline-range compounds. It is therefore the most desirable fluid system to encounter in a reservoir.

RETROGRADE CONDENSATE GAS

MIDDLE 50's < °API < 70
CONDENSATE μ ≈ 0.25 CP
2000 TO 6000 < R INITIAL < 15000

The hydrocarbons in the reservoir are gaseous at initial conditions. During initial gas production, large quantities of condensate drop out in the surface facilities. Later on, as reservoir pressure declines, the condensate begins to drop out in the formation. "Retrograde" refers to the fact that these gases liquefy when pressure is reduced, while normal fluid behavior is for gas to liquefy when pressure is increased.

WET GAS

CONDENSATE °API > 60
CONDENSATE μ ≈ 0.25 CP
2000 TO 6000 < R INITIAL < 15000

All hydrocarbons in a wet gas reservoir are in a gaseous state. When the gas is produced, the associated cooling causes condensate to drop out in the surface facilities.

DRY GAS

The gas is composed primarily of methane with only small amounts of ethane, propane and butane. These reservoirs produce no condensate.

Oil Reservoirs – Primary Drives

The reservoir drives discussed next are effective only in low shrinkage and high shrinkage fluid systems; that is, reservoirs with gravity > 20° API. Heavy oil, tar, and bitumen are essentially unproducible under primary conditions. Reservoirs are usually subject to more than one drive mechanism. To aid understanding, however, this discussion deals with each drive in isolation.

Solution gas drive.   Solution gas is the principal drive mechanism in perhaps a third of the world's reservoirs and contributes significantly in many more. It is therefore the most widespread reservoir drive.

A purely solution-gas drive reservoir has no gas cap, so its pressure is greater than the bubble point. There is no water influx, nor is the volume of bottom-water sufficient to support a water-expansion drive.

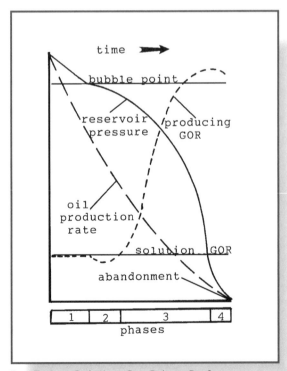

FIG. 5–1   Solution Gas Driven Performance

With reference to Figure 5-1, solution gas drive performance has four phases from initial production to abandonment:

Phase 1: When production starts, the reservoir pressure declines rapidly to the bubble point. All the gas is still in solution in the oil so the producing GOR is the same as the solution GOR.

Phase 2: When pressure falls to the bubble point, the producing GOR dips below the solution GOR for a short period. This is because some gas comes out of solution, but is not produced because the critical gas saturation is not yet reached. The gas remaining in solution is the only gas that is produced. Expansion of the free gas slows the rate of pressure decline.

Phase 3: The critical gas saturation is exceeded and free gas is produced along with the oil and its remaining solution gas. Producing GOR increases as decreasing pressure liberates more solution gas.

Gas expansion from the continuing pressure drop is the principal drive mechanism. Oil expansion also contributes.

The rate of decline of oil production is slowed by the slower pressure decline but accelerated by the relative permeability effects of the increasing gas saturation (accompanied by decreasing oil saturation).

In the later stages of phase 3, the pressure drops rapidly because of the high gas production rate.

Phase 4: The producing GOR turns downward, the remaining oil is dead (depleted of its solution gas) and production declines to its economic limit.

The recovery factor for solution gas drives worldwide ranges from 5 to 30%, with an average of 15 to 17%. This low recovery with at least 70% of the oil remaining after primary recovery makes solution gas drive reservoirs good candidates for secondary recovery.

Water drive. Water-drive reservoirs have access to aquifers that provide water to replace some or all of the volume of fluids produced. This water influx maintains reservoir pressure wholly or partially, resulting in higher recovery factors.

Recovery factors for water drive reservoirs can be as high as 70%, and average 35% to 40%. No other primary oil drive provides recoveries this high. Roughly a third of the world's reservoirs have water drives (FIGURE 5-2).

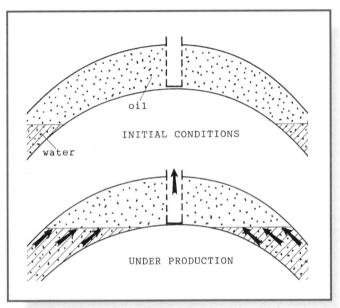

FIG. 5–2   Water Drive Reservoir

There are two possible sources of water for water drives (FIGURE 5-3):

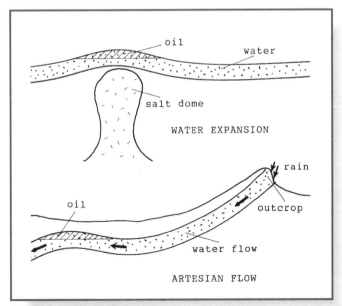

FIG. 5–3   Water Drives

1. Water expansion caused by reservoir pressure decline can provide enough water if the portion of the reservoir that is water-saturated is many times larger than the oil-saturated volume.

2. Artesian flow can occur if the reservoir formation outcrops in nearby mountains.

The rate of water influx varies greatly between reservoirs. In partial water drives, pressure declines unless production is sharply restricted. Solution gas is the dominant drive mechanism. With active water drives, however, there is adequate water available to maintain the pressure with unrestricted production. This performance is shown in Figure 5-4.

Active water drive performance:

- The pressure decline is minimal because incoming water completely replaces the produced fluids.

- Water production starts at some point. The water cut (percent of total liquid production) increases until abandonment, when it often exceeds 95%.

- The producing GOR stays constant since the bubble point pressure is never reached.

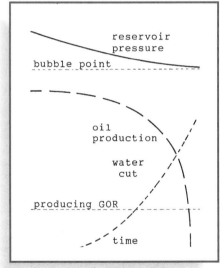

FIG. 5–4   Active Water Drive Performance

- Oil production initially holds steady because of the minimal pressure drop. When water production starts, however, the water displaces some of the oil being produced.

Gas-cap drive

As oil is produced from a gas-cap reservoir, the gas in the cap expands and prevents the pressure from dropping as rapidly as in a solution gas drive. Gas-cap recovery factors are therefore higher, averaging 30% to 40% (FIGURE 5-5).

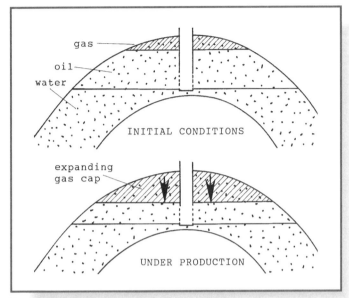

FIG. 5–5   Gas Cap Drive

When a gas cap is present, it's usually better to produce only oil initially. As the reservoir pressure drops from oil production, the gas cap expands, retarding the rate of pressure decline and increasing oil recovery. Only after the oil leg is depleted is the gas cap blown down.

Gas caps are initially at bubble point pressure. However, if a gas cap's volume is many times the size of the oil zone (a somewhat unusual occurrence), the oil zone may be depleted without the pressure declining enough for free gas to break out in the oil zone.

With a smaller gas cap, however, the reservoir pressure drops significantly as oil is produced and free gas breaks out, reaches its critical saturation and is produced with the oil. As with solution gas drive, this reduces the relative permeability to oil and impairs recovery.

The effectiveness of a gas cap drive therefore depends on its size relative to the size of the oil zone.

Gravity drainage

After pressure depletion of reservoirs with very high vertical permeability, the remaining oil and gas may separate with the oil pooling in the bottom of the reservoir. It is occasionally profitable to produce this oil, but gravity drainage does not make a significant contribution to the world's reserves.

Compaction drive

Overpressured reservoirs may experience compaction as they deplete because the overpressured fluids had prevented normal compaction. The North Sea chalk fields are an example of this.

Gas Reservoirs

Wet and dry gas reservoirs

The reservoir mechanics of wet and dry gas reservoirs are identical because in both cases there are no liquid hydrocarbons in the formation.

Gas has an extremely low viscosity (less than a hundredth of most oil viscosities), so it flows readily through the formation to the wellbore. Gas wells are therefore drilled on much wider spacing (distance between the wells) than oil wells.

The operation of the gas-expansion drive mechanism could not be simpler. As reservoir pressure drops, the gas expands into the wellbore. This continues until the reservoir pressure is essentially zero. In some cases, compressors are installed to suck out the last few MCF. Recovery factors often reach 95%.

Unlike oil, gas is highly compressible so its surface volume is much greater than its reservoir volume. The ideal gas law states that the SCF of gas occupying a given volume of reservoir space is directly proportional to the pressure. For example, a cubic foot of gas volume at 5000 psi in the reservoir would actually contain 340 (5000÷14.7) standard cubic feet of gas.

Not all gas reservoirs are constant volume. Active water drives are common but are considered a negative because the encroaching water usually bypasses considerable gas. Compaction drives also occur.

Retrograde condensate gas reservoirs

Production from retrograde condensate reservoirs is initially very profitable because of the rich condensate yields of two or three hundred barrels per

MMCF. Problems start, however, when the reservoir pressure drops to its dew point. The condensate then starts dropping out in the formation. This is a two-pronged problem. First, the operator loses most of the revenue from condensate sales. Second, and more serious, is that the liquids building up in the formation restrict its permeability to gas. As a result, gas production drops off sharply. Quite suddenly, this reservoir is not nearly as profitable as it first appeared to be.

One remedy is to initiate a gas-cycling project. This entails stripping the natural gas liquids from the produced gas stream, then compressing and re-injecting the residue gas. The dry gas circulates through the formation and re-vaporizes some of the liquids which are then stripped out at the surface. Carried to its conclusion, cycling can result in very high recoveries.

The distressing reality of gas-cycling is, however, that if there is a market for the residue gas—no matter how low-priced—the time-value-of-money renders cycling uneconomic. The present value of the reservoir is optimized by selling off the residue and then abandoning the reservoir that is loaded up (rendered unproducable) with liquids.

Retrograde condensate reservoirs are therefore candidates for recycling only if discovered in areas remote from markets.

Waterflooding

Waterflooding is usually a secondary drive mechanism initiated after the primary depletion of solution-gas drive reservoirs. A rule-of-thumb is that waterflooding will recover half-again as much oil as was produced under primary.

In recent years, however, waterflooding has increasingly been initiated prior to primary depletion, often simultaneously with the start of primary depletion. This "pressure maintenance" approach is not only more profitable on a time-value-of-money basis, but also results in higher recovery.

Waterflooding is the dominant improved recovery mechanism throughout the world with thousands of active projects. It is said to have gotten started accidentally when salt water entered wells through casing leaks, dumped downhole into depleted producing zones and stimulated production in surrounding wells. Many "dump floods" were then intentionally started by perforating uphole saltwater zones (FIGURE 5-6).

Waterflooding's advantages.
Waterflooding's many advantages explain its popularity.

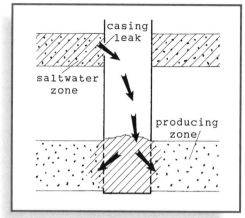

FIG. 5-6   Dump Flood

- Saltwater is readily available and inexpensive. Most fields produce significant volumes of saltwater that need to be disposed of, so waterflooding turns disposal cost into a benefit. The necessary "make-up" water (water needed in addition to produced water) can usually be obtained from saltwater zones downhole close to the pay zone.

- Since produced saltwater occurs naturally in the producing zone, it is chemically compatible with the reservoir fluids. If waters from other sources are used, incompatibility may cause precipitation of solids that plug the formation.

- Waterflooding equipment is relatively simple and inexpensive. Although saltwater is corrosive, stainless steel valve trims, plastic coatings for pipe, etc. are able to handle it at reasonable cost.

- Waterflooding is a safe operation because saltwater is non-toxic and non-flammable and pressures are relatively low.

- The biggest advantage of waterflooding is simply that it works so well. Very few floods have turned out to be uneconomic.

Fluid displacement. Saltwater and oil do not mix, so waterflooding is an immiscible displacement process.

Wettability. Because they were deposited in saltwater, the individual sand grains in most sandstone reservoirs are preferentially water-wet. This means that water, not oil, adheres tightly to their surfaces. The effects of wettability are clearly seen in waxing a car. When a car's finish needs waxing, rainwater spreads itself thinly over the surface. This attraction between the surface and the water indicates the surface is water-wet. After waxing, however, the rainwater pulls itself away from the surface and stands up in beads on the car's surface. The surface is now repelling the water because the petroleum wax has made it oil-wet.

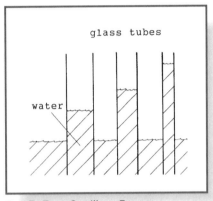

FIG. 5–7 Capillary Forces

Capillary forces. The attraction between a surface and its wetting fluid is the basis of capillary forces, which are an important drive mechanism in waterflooding (FIGURE 5-7).

Capillary forces are evident when glass tubes are placed in a bowl of water. Since clean glass is water wet, the attraction of the glass pulls the water upward all around the inside of the tube, supporting a higher water level in the center of the tube. Since smaller tubes contain smaller, lighter columns of water relative to the length of the wetting surface, they support higher fluid columns. The tiny pores in water-wet reservoir rocks therefore act like capillary tubes and draw the water through the formation.

Although the pressure imposed by the injection-water pump is the principal fluid displacement mechanism, capillary forces are clearly a significant adjunct. They are uniquely effective in displacing oil from the networks of tiny pores that would be bypassed by pressure alone.

Sweep efficiency. Waterflood recoveries depend not only on the intergranular displacement discussed above, but also on how much of the reservoir rock is actually contacted by the advancing water (sweep efficiency).

Waterfloods are often visualized with a vertical flood front building up a pressurized oil bank and sweeping piston-like from a water-injection well to a producing well . This is a serious over-simplification.

Mobility ratio. A fluid's mobility is its ease of movement through the reservoir. The mobility ratio in a waterflood is the mobility of the displacing fluid (water) divided by the mobility of the displaced fluid (oil).

A flood's sweep efficiency is effected by its mobility ratio. If, for example, a thick viscous oil is being swept by far less viscous water, the water tends to break through and establish channels, bypassing much of the oil. In the reverse situation where a light, low-viscosity oil is being displaced by water thickened with a polymer, the water tends to spread out over a broad front, resulting in a high sweep efficiency. A mobility ratio of one or less is generally considered to be good for a flood's sweep efficiency, while ratios greater than one are detrimental.

Viscosity is a major determinant of mobility ratio (along with relative permeability), so reservoirs with heavy, viscous oil generally flood poorly, while reservoirs with high-gravity, low-viscosity oil flood more efficiently.

Reservoir heterogeneity. There is a tendency to visualize sandstone reservoirs as having constant characteristics throughout. This is emphatically not the case. The natural forces that formed the reservoir were in a constant state of flux, with the result that heterogeneities abound both vertically and horizontally.

Perhaps the most common heterogeneity encountered are barriers to vertical permeability. A momentary (in geologic time) change in depositional conditions often leaves almost invisible horizontal layers of clay within otherwise homogeneous sandstones. Instead of having one thick reservoir, the clays restrict vertical permeability, forming multiple, thin reservoirs.

Carbonate reservoirs have even greater permeability variation that sandstones. The formation of dolomitic porosity, for example, is highly erratic as is the deposition of calcium carbonate in existing porosity.

The various heterogeneities cause most reservoirs to waterflood unevenly with high-permeability thief zones taking more than their share of the water. This leads to early breakthrough of water into producing wells. Figure 5–8 presents this more realistic view.

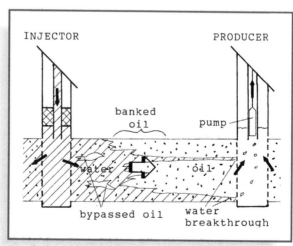

FIG. 5–8   Waterflood

In practice, a great deal of oil is bypassed and early water breakthrough usually occurs, but the flood continues to operate profitably. The breakthrough channel becomes a low pressure conduit for a high volume of water, so it's often necessary to increase the water injection rate to keep pressure on the unswept formation. Larger pumping equipment is installed to handle the water. The water-oil ratio (WOR) of floods often reach 95% or higher before they become uneconomic.

Well patterns. In the 1950s and 60s when waterflooding was relatively new, a number of peripheral floods were tried. The plan was for the injected water to gently encroach from the perimeters of the field, much like a water drive, and result in maximum recovery. A few injection wells were therefore located around the edges of the reservoir (FIGURE 5-9).

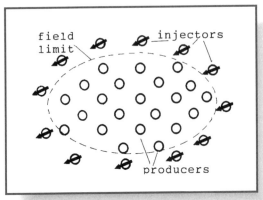

FIG. 5–9   Peripheral Flood Pattern

It soon became apparent, however, that with injection rates limited by so few injection wells, the projects would take too long. From a time-value-of-money perspective, this was not optimizing present worth.

The industry now recognizes that a much higher ratio of injection wells to producing wells is needed. In fact, the current practice is generally one injector for each producer. The wells are normally arranged in a five-spot pattern which has alternating injectors and producers. If the field was originally depleted under primary conditions, the injectors are converted producers (FIGURE 5-10).

Each individual five-spot pattern has five wells—one

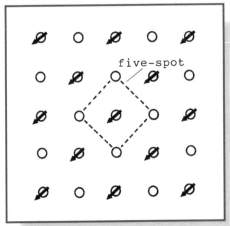

FIG. 5–10   Five-spot Flood Pattern

injector and four producers (or vice-versa). The injected water front moves preferentially in the direction of the pressure drawdowns created by the producers rather than outward in a strictly radial pattern. This results in a large portion of the reservoir—perhaps 50%—being bypassed (FIGURE 5-11).

Combining the oil that is bypassed in the vertical dimension as shown in Figure 5-10 with the oil that is bypassed in the horizontal dimension as shown in Figure 5-11, suggests less than half of the reservoir is contacted by the flood water. These pockets of bypassed oil have been pressured by the flood and can often be produced profitably by infill drilling at the spots indicated in Figure 5-11.

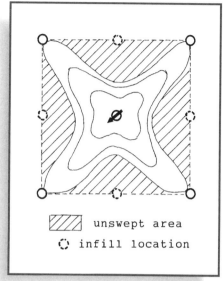

unswept area

infill location

FIG. 5–11   Five-spot Sweep Pattern

Thermal recovery

The viscosity of heavy oil ($< 20°$ API) is so high that the oil generally does not move through the formation and enter the wellbore at commercial rates. Thermal recovery techniques are therefore used to inject heat into the formation, which reduces the oil's viscosity and results in a higher producing rate.

Fire flooding (*in situ* combustion). A compressor at the surface injects air down injection wells, providing the oxygen necessary to ignite the oil in the formation. The hot gases from combustion penetrate out into the formation, warming the crude and reducing its viscosity.

Fire flooding has been tried in many places, but has had few outright successes. The problem is that the cold oil remaining out ahead of the warming gases is very persistent in resisting flow from the injector to the producer. The only way to get movement is to drill the wells very close together, which usually renders the project uneconomic (FIGURE 5-12).

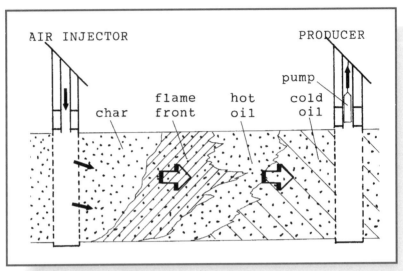

Fig. 5–12   Fire (in situ) Flood

Steam flooding. Steam is generated at the surface and injected down wells to the formation face. As the heavy crude is heated, the steam cools and condenses into hot water (Figure 5-13).

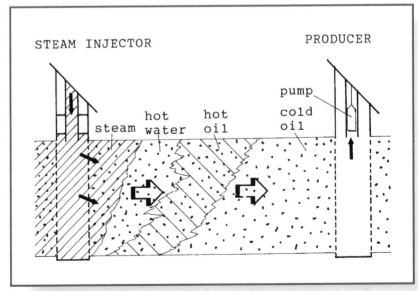

Fig. 5–13   Steam Flood

Steam flooding has the same problem as fire flooding with respect to the immobility of the remaining cold oil. There are, nevertheless, many profitable applications for steam flooding. For example, the heavy oil production in southern California is profitably steam flooded because it is so shallow (200-500 feet). Wells are cheap to drill at that depth, so these fields are developed with spacing as close as two and one-half acres per well. With the wells this close together, the cold oil can be moved.

Cyclic steam flooding (huff and puff). This is the most successful thermal recovery mechanism. It is successful because it does not attempt to move cold oil. Instead, each well is independently cycled as follows (FIGURE 5-14):

1. Steam is injected into the well for a week or two, heating the oil in the immediate vicinity of the wellbore.

2. The well is soaked for a few days to allow the heat to disperse.

3. The pumping equipment is activated and the reduced viscosity oil in the vicinity of wellbore is produced for a week or two.

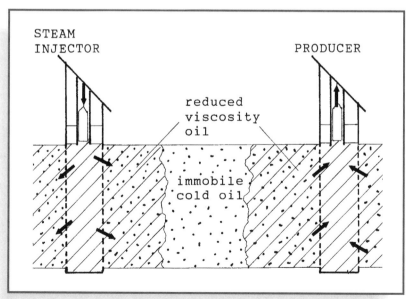

FIG. 5–14   Cyclic Steam Flooding (huff and puff)

In field operations, two wells typically share a common steam line. Both are equipped with sucker-rod pumping systems. While steam is injected into one well, the second is pumping. Injection stops for the soak period, then steam is switched to the second well and the pumping unit is started on the first well.

In summary, the advantages of cyclic steam flooding are that

- It requires modest capital investment, is operationally simple and inexpensive and, unlike steam flooding, does not require drilling large numbers of tightly spaced wells.

- It sharply stimulates production. Equally important because of the time value of money, the production (and revenue) increases immediately.

- Operations can profitably continue for many decades.

Because of these significant advantages, cyclic steam flooding has broad application throughout the world.

Miscible flooding

Oil recoveries are improved when a reservoir is flooded with fluids that are miscible (mix with) the reservoir oil. By eliminating the interfacial tension that holds immiscible fluids apart, the miscible flood front is able to penetrate deeply and displace fluid from the most remote recesses of the rock matrix.

Carbon dioxide flooding is the only miscible process currently undergoing much growth. Other processes, while effectively improving recovery, are generally too expensive.

Carbon dioxide flooding. Carbon dioxide gas (CO_2) occurs naturally in underground reservoirs, often in conjunction with hydrocarbon gases. Its distribution around the world is limited, but when it is available in the vicinity of oil fields, it can be a very effective flooding agent.

CO_2 is usually applied as a tertiary process following primary depletion and waterflooding. The center of CO_2 activity is West Texas in the United States, but there are projects operating in Europe, Turkey, and elsewhere.

Process description: CO_2 is transported to the wellhead by pipeline and injected under pressure. To assure miscibility, the reservoir must be maintained at a relatively high pressure (FIGURE 5-15).

The CO_2 combines with the oil, swelling the oil and reducing its viscosity. CO_2 injection is then usually stopped and a slug of salt water is injected. The flood then proceeds with alternating slugs of CO_2 and water sweeping toward the producing well. This is called the WAG (water and gas) process which is designed to

- Counter CO_2's tendency to rise to the top of the formation and prematurely break through to the producing well. The slugs of heavier water tend to maintain a more vertical flood front.

- Improve sweep efficiency because water's lower mobility ratio encourages a broad, even advance of the flood front.

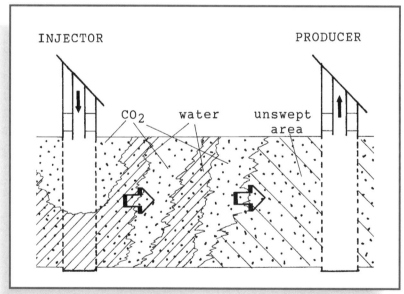

FIG. 5–15 CO_2 WAG Flood

A major portion of the project's capital investment is for surface treating facilities to separate the CO_2 from the hydrocarbons in produced gases. The CO_2 is then compressed and re-injected.

Other miscible processes. The following miscible processes have all been tried and found to be workable. New projects are, however, rarely initiated because of economics and supply considerations.

Liquefied petroleum gases (LPG). LPG is readily miscible in crude oil, and when followed by a water sweep can result in high recovery. Since much of the LPG is not recovered, however, the process is too costly when LPG prices are strong.

Flue gases. Waste gases from burning oil or gas in large industrial applications will attain miscibility in the reservoir. This is not a significant process because of the limited availability of flue gases.

Nitrogen. Nitrogen extracted from air (26% nitrogen) will attain miscibility, but only at pressures so high that only a few of the deepest reservoirs are candidates.

Detergents. Detergents injected into a formation have an effect very similar to miscibility. When followed by a water sweep, high recoveries can be attained. The cost of detergent, however, is too great for the process to be economic.

Alkaline (caustic). When caustics are injected, they react chemically with some crudes to form detergents. They have the same pluses and minuses as detergents, with the added problem that they are unpleasant and unsafe to handle.

Mobility-ratio improvement

As discussed under waterflooding, a low mobility ratio results in a high sweep efficiency. A great deal of work has been done on enhancing waterflood mobility ratios by using polymers to thicken an initial slug of injection water. The slug is followed by ordinary saltwater.

Polymers and other mobility-ratio enhancers are still of interest with ongoing research and field trials, but the high cost of the chemicals generally limits its commercial application.

Microbial floods

The idea of using microbes to enhance oil recovery is still very much in the research stage, but it should not be dismissed too lightly.

The original concept was to introduce crude-eating bacteria at the injection wells. Several species were available for experimentation. As the bacteria consumed oil, their mass would grow and force the oil towards the producing wells. Their gaseous waste products would enhance the drive by reducing the oil's viscosity.

The current status is that the growth of the crude-eating bacteria was found to be too slow. Experimentation is now focused on strains that are fed with molasses or other nutrients pumped down the injectors. Some encouraging results have been obtained from pilot projects.

Oil and Gas Reserves

Reserves, as defined by the Society of Petroleum Evaluation Engineers (SPEE), are estimated volumes of crude oil, condensate, natural gas, natural gas liquids, and associated substances anticipated to be commercially recoverable from known accumulations from a given date forward, under existing economic conditions, by established operating practices, and under current government regulations.

Some essential points regarding reserves:

- Reserves have not yet been produced—they are still underground.

- Being underground, reserves cannot be accurately counted. Estimates of their volume entail numerous assumptions and considerable uncertainty.

- A field's reserves cannot be delivered at a single time since it takes years to deplete the field. Their sales price is therefore impossible to predict.

- Reserves are the major wealth of upstream oil companies.

Given the importance and the ambiguity of reserve estimates, oil companies carefully apply conservative guidelines developed by the SPEE and the Society of Petroleum Engineers (SPE). Elements of these guidelines follow.

Reserve Categories

Proved Reserves

- Can be estimated with reasonable certainty

- Generally must be supported by actual production or formation tests. Indirect indications like electric logs are usually not adequate

- Include both drilled locations and the undrilled locations that directly offset commercially producing locations

- Must have operational facilities to process and transport the reserves to market

- Have publicly–quoted company reserve statistics

Unproved Reserves

- Are less certain than proved reserves

- May require future economic conditions different from current conditions

- Are estimated for internal planning or special evaluations, but not routinely compiled

- May be divided into two risk classifications: probable and possible

Reserve Determination Methods

Early in a field's life, little is known about the extent, quality, and drive mechanism of the reservoir. Reserve estimates are therefore likely to be highly inaccurate—either on the plus or minus side. As the field is developed, and particularly as data on production performance accumulates, the range of error narrows. It is, however, not until the last barrel is produced and the field abandoned that the reserves are known precisely.

Listed below are various methods that are used either singly or in combination to estimate reserves.

Volumetric method. The volumetric method is primarily used early in the field's life when the rough dimensions of the reservoir are known but performance data is not yet available. It is a relatively crude method, subject to significant error.

Step 1:   The original oil in place is calculated by the formula

$$OOIP = 7758 \times A \times H \times \Phi \times S_0$$

where   OOIP = Original Oil In Place, bbls
7758 = Bbls / acre-foot constant
A = Reservoir Area , acres
H = Reservoir Thickness, feet
Φ = Porosity, decimal
S_0 = Oil saturation, decimal

Step 2:   The barrels of recoverable oil is calculated by the formula

$$STB = \frac{OOIP \times RF}{\beta}$$

where   STB = Stock Tank Barrels

RF = Recovery Factor (e.g., 35%) reflecting the fraction of OOIP similar fields in the vicinity have recovered

β = Reservoir volume factor reflecting shrinkage the oil undergoes as it is produced

Step 3:   Associated gas reserves are calculated by multiplying the oil reserves by the solution Gas-Oil-Ratio (GOR)

Material balance method. After approximately 10% of the reserves are produced—enough to measurably reduce the reservoir pressure—the material balance method yields greater accuracy than the volumetric method.

The reservoir is viewed as a constant-volume tank from which known volumes have been extracted. Having analyzed reservoir samples to gain knowledge of the pressure-temperature-volume behavior of the hydrocarbon system, the remaining reserves can be calculated from the observed reservoir pressure. If water influx is occurring, adjustments can be made to account for it.

Reservoir modeling / simulation. When a great deal of high-quality data is available on a reservoir, a more sophisticated material balance technique requiring computer solution can be applied. A multiple-cell computer model of the reservoir is created and the depletion process is mathematically simulated. Under the right conditions, these simulations can be quite reliable.

Production decline curve method. As a reservoir reaches a more advanced state of depletion, its production rate typically declines along a predictable path. In many cases, the decline is close to exponential; that is, at a constant percent. When the rate is plotted on a logarithmic scale versus time on a Cartesian scale, the decline becomes a relatively straight line which can be extrapolated to yield reserves and a forecast of future production.

The extrapolation ends when the line reaches the economic limit—a horizontal line on the graph. This represents the break-even point where revenue equals direct operating expenses and the project is abandoned.

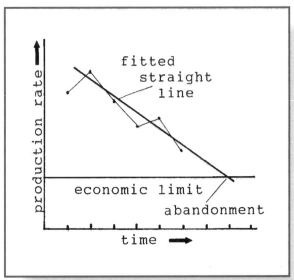

FIG. 5–16   Reserves Estimation by Decline Curve Extrapolation

This method is effective for one well, a group of wells, or the entire field and is considerably more accurate than other methods. It therefore dominates reserves estimation. A number of excellent computer programs are available that display the decline curves, facilitate curve fitting, and calculate the results.

Reserve revisions

When new information becomes available, companies revise their reserve estimates. For example:

- Completion of successful new wells extend the reservoir's limits; dry holes contract the limits.

- Movement in crude oil or gas prices changes the economic limit with higher prices increasing reserves and lower prices decreasing reserves.

- Initiation of a waterflood or other enhanced recovery project increases reserves.

- New evidence can change reservoir drive assumptions. For example, if it becomes apparent that an active water drive is present, the recovery factor may be increased.

Companies also routinely review the reserve estimates on all their fields once a year for inclusion in the annual report.

Reserves are not assets

Because of the uncertainties in both the counting and the pricing of reserves while they are still in the ground, this major segment of upstream oil companies' worth is not included as an asset on the balance sheet.

In the 1970s, an experiment in Reserves Recognition Accounting was tried in the United States. The value of oil companies' reserves was estimated and published in an appendix to their annual reports, with the idea that the reserves would be promoted to assets the following year. However, the effort proved so traumatic to accountants that the whole idea was dropped.

The present situation is that proved reserve estimates, valued at current prices, are listed in the annual reports, but reserves are not counted as assets on the balance sheet. It is therefore essential to realize that oil company asset values are not on a comparable basis with other types of companies.

DRILLING 6

Introduction

The mechanics of rotary drilling are quite simple. A rotating bit breaks loose the rock at the bottom of the hole. The rock fragments are then swept away and lifted out of the hole by the mud stream. The practice of rotary drilling is, however, extremely demanding and complex. Some of the reasons for this are

- The difficulty of working at the bottom of a hole three or four miles deep
- Heavy, complex equipment under stress
- Variable rock characteristics
- Reservoir fluids—oil, water, and gas—entering the hole
- Remote areas and hostile environments
- Non–stop operations—24 hours per day, 7 days per week
- Dangerous working conditions
- Environmental hazards
- Pressure to hold down costs

Early Drilling Rigs

Until early this century, oil and gas wells were drilled with cable–tool rigs. The hole was made by spudding, which is making a hole by reciprocating a heavy, chisel–shaped bit suspended on a hemp or steel line (FIGURE 6–1).

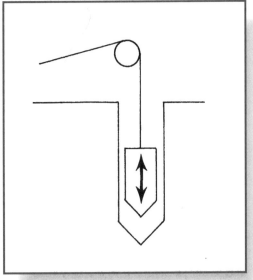

FIG. 6–1   Spudding

The major problem with cable–tools is that there was nothing to prevent oil and gas from blowing out at the surface whenever a reservoir was penetrated. This is wasteful and dangerous, and today would have major environmental repercussions. The rotary drilling system in use today avoids this problem by keeping the hole full of heavy mud that holds back reservoir fluids.

Rotary Drilling System

Drillbits

The only useful work done by a drilling rig takes place at the bottom of the hole where the bit drills the rock. Everything else on a rig—the prime mover, derrick, drawworks, drill pipe, etc.—simply support the bit in this effort.

Drillbits operate in an extremely dirty and hot environment, frequently bounce up and down, and in general are subjected to enormous mechanical stresses. To replace a bit that has become dull or has broken is costly. The entire drill string must be pulled, the bit changed, and the drillstring re-run. In deep holes, this roundtripping can take hours of rig time. The hole can also be lost from pieces of hardened steel left in the hole from a broken bit. It's therefore essential that bits be extraordinarily competent and long-lasting. Fortunately, they are exactly that.

Tricone roller bits. These are the most commonly used bits. Clockwise rotation of the drillstring causes rotation of the three roller–cones, bringing successive teeth to bear vertically on the bottom of the hole. The weight concentrates at the point of the teeth and crushes the rock. The fragments (cuttings) are then swept away by mud jetting out of three downward–facing nozzles (FIGURE 6–2).

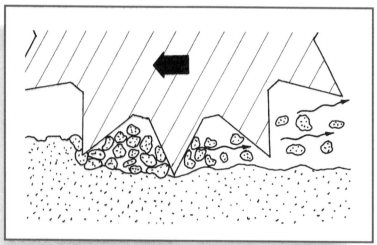

FIG. 6–2   Rock Crushing with Roller Bit

Tricone bits can either have teeth milled from hardened steel as shown in Figure 6–2 or, for harder rock, smaller tungsten carbide teeth inserted in a steel cone, shown in Figure 6–3. When the points of the teeth wear down, the drilling rate slows and the bit must be replaced. Bearing wear can also be a problem.

FIG. 6–3   Tricone Roller Bit with
             Tungsten Carbide
             Insert Teeth

PDC (polycrystalline diamond cutter) bits.   This is a relatively new bit design that uses very hard polycrystalline diamond wafers on the face of protruding bronze bases (FIGURE 6–4). These are drag bits, which means they have no moving parts. They drill with a shearing action that is particularly effectively on the relatively soft formations found in coastal and offshore areas (FIGURE 6–5).

FIG. 6–4   PDC Bits and One Diamond Bit

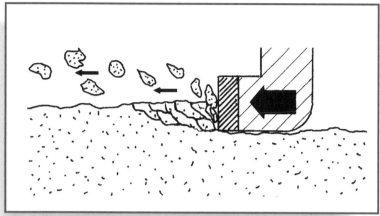

FIG. 6–5   Shearing Action with a PDC Bit

Drillstring

The drillstring is a tube of steel pipe extending from the surface down to the bit. It serves to

- Transmit rotational energy from the surface to the bit
- Conduct the mud stream from the surface to the bit
- Put weight on the bit to maximize penetration rate

The drillstring is made up of 30–foot joints of steel pipe, screwed together. The specialized couplings , called tool joints, have coarse, tapered threads designed to withstand repeated break–outs (unscrewing) and make–ups (screwing together) while retaining tensile strength and sealing ability. Three different pipe configurations are used in the drillstring (FIGURE 6–6):

- *Drill pipe* makes up the bulk of the string. The tool joint is larger in diameter (upset) than is the shank of the joint.

- *Drill collars* are thick–walled joints of drill pipe used to put weight on the bit. Several collars are normally run at the bottom of the string, immediately above the bit. The tool joints are the same diameter (not upset) as the shanks of collars.

- The *kelly* is the topmost joint of drillpipe in the drillstring. It has flattened sides (four or six) used to transfer torque from the rotary system.

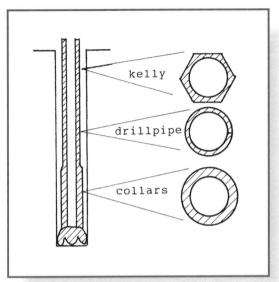

FIG. 6–6   Drillstring Configuration

Hoisting system

By far the heaviest work on a drilling rig is performed by the hoisting system as it runs the drillstring in and out of the hole. The components of the system are:

- The *derrick*, which is tall enough to pull triples (a three–joint "stand" of drillpipe) and strong enough to bear the weight of the entire drillstring.

- The *drilling line* of steel wire rope.

- The *drawworks*, with a large rotating drum that spools in drilling line to raise the load or spools out line to lower the load.

- The *crown block* mounted in the top of the derrick and the traveling block connected to the crown block by the drilling line. Each block is an assembly of multiple pulleys (called "sheaves," but pronounced "shivs") that the drilling line is wrapped around. Taken together, the blocks and drilling line constitute a block and tackle—a device to multiply the lifting capacity of the drilling line. The eight wraps of the line shown in Figure 6–7 permit eight times the load to be lifted as would be possible with the single line. This mechanical advantage does, however, have a cost. For every foot the traveling block is raised, eight feet of line has to be spooled onto the drum.

• The *drilling hook*, which is attached to the traveling block and supports the load.

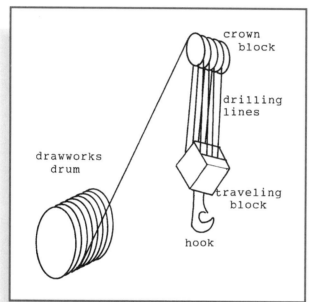

FIG. 6–7 Drilling Rig Hoisting System

Rotating systems

There are three alternative systems used to rotate the drillbit:

1. The *conventional rotating system* is used with most onshore wells and many offshore wells (FIGURE 6–8). The kelly joint, with the rest of the drillstring suspended below it, passes through the kelly bushing and attaches to the swivel hanging from the drilling hook. The rotary table spins the kelly bushing, whose flat–sided hole matches the flat sides of the kelly. This rotates the drillstring and bit.

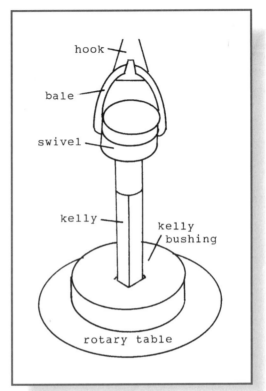

FIG. 6–8   Conventional Rotating System

2. The **power swivel**, or top–drive, is relatively new and finding increasing application, particularly offshore. It uses a hydraulic motor integrated into the swivel to transmit the rotary power. This replaces the kelly, kelly bushing and rotary table. Top–drives have the advantages of permitting circulation and rotation while pulling out of the hole and of adding a full stand of pipe (3 joints) rather than only one joint at a time as the hole deepens.

3. The **down–hole mud motor** is used for directional drilling. The motor is mounted directly above the bit, and is powered by the drilling mud. It turns the bit without rotation of the drillstring.

    Mud motors are normally used in the deviated portion of the hole only, with the vertical section drilled conventionally by rotating the drillstring.

Drilling fluids (mud)

In rotary drilling, mud is continuously circulated down the drillstring, out through the nozzles in the bit, then back to the surface through the annulus (space between the drillstring and the hole). The most commonly used mud system is saltwater with numerous additives. Oil–based mud (water in oil emulsion) is also used where its greater stability justifies its greater cost. Various "synthetic" chemical muds, which are extremely expensive, are also used. Their application is offshore wells that require the stability of oil–base mud, but where its use is prevented because disposal of the drill cuttings overboard would cause pollution.

Functions of drilling mud:

- The mud jets forcefully out through the bit nozzles, flushing the drill cuttings away from the bit face. This cleans the rock surface, increasing the bit's penetration rate.

- The mud carries the cuttings up the annulus and out of the hole. To give saltwater mud the necessary lifting ability, it is thickened with clay, called gel.

- The mud lubricates and cools the bit, greatly extending the bit's useful life.

- The mud exerts backpressure on the exposed formations to prevent influx of formation fluids, called a kick, which could lead to a blowout. The density of the mud used therefore depends on the formation pressures anticipated. The density of saltwater mud is increased as needed (called weighting up) by adding pulverized barium sulfate (barite).

- The mud seals off permeable formations to prevent mud loss. In extreme cases such as fractured zones, the formation may absorb the entire mud stream, resulting in lost circulation (no mud returns) which could lead to a blowout. Solid materials, such as shredded paper or cottonseed hulls, are added to the mud to prevent lost circulation.

- The mud reduces friction between the drillstring and the side of the hole.

Mud engineers monitor the mud's characteristics, making changes as needed. For example, if additional gel strength is needed, sacks of dried gel is mixed into the mud stream. If additional weight is needed, sacks of barite or other weight material is added. Since higher gel strength and weight slow the bit's penetration, saltwater mud is frequently diluted with water to maintain the absolute minimum gel strength and weight necessary. It's also common to drill "underbalanced" (having a lighter mud column than would be necessary to hold back reservoir fluids) in situations where there is no possibility of a kick.

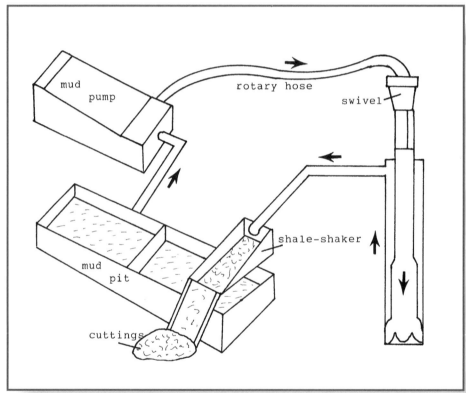

FIG. 6–9   Mud Circulation System

Mud circulation system (FIGURE 6–9):

1. As it comes out of the hole, the mud is diverted to the side before it reaches the rig floor.

2. It falls down through the shale–shaker, which has calibrated mesh that strains out the large cuttings.

3. The mud then drops into the mud pit. Pits are normally constructed of steel, but earthen pits with plastic liners are often
   used onshore.

   Fine–grained solids settle out in the pit. In some situations,
   mechanical desilters and desanders are used.

4. The mud pump then picks up the solid–free mud out of the far
   end of the pit.

5. The mud is pumped through the flexible rotary hose to the
   swivel, then down the drillstring and back up the annulus to
   complete the circuit.

Pipe handling

The following devices are used to handle the pipe going in and coming out of
the hole:

- *Elevators* attached to the hook are used to lift the pipe string.
  They wrap around the pipe below the tool joint and latch, then
  lift against the upset shoulder.

- *Slips* are used to hang the string from the rig floor. They wrap
  around the pipe, wedging it against a recess in the floor. When
  the string is picked up, the slips come free and can be
  removed. Slips are normally operated by hand, but power slips
  are also available (FIGURE 6–10).

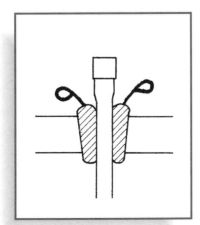

FIG. **6–10**   Slips

- *Tongs* are used to make–up and break–out the pipe joints. Power tongs spin the pipe; back–up tongs keep the rest of the string from turning. At times the rotary table is used to spin the pipe.

- The *fingerboard* in the derrick has protruding fingers that hold stacked stands of pipe in place .

- The *rathole* is a hole in the rig floor where the kelly and swivel are stored when they are not in use.

- The *mousehole* is a hole in the rig floor where a joint of pipe is placed prior to being added to the string.

Prime mover

The prime mover is the source of power for the entire rig. Most modern rigs are diesel–electric. Diesel engines drive electric generators that produce direct current to power all the rig's equipment.

Routine Drilling Procedures

Drilling ahead

Most of the rig's time is occupied by drilling ahead, also called "turning to the right" or "making hole." The bit is on bottom drilling, and this continues until the kelly joint is drilled down (upper coupling approaching the rig floor). Another joint of drillpipe is then added, called making a connection, and drilling resumes. When a top drive is used, the entire stand (90') is drilled down, and an entire new stand then added.

While drilling ahead, the driller watches the weight indicator closely, slacking off on the brake to lower the drillstring and keep the proper weight on the bit. He is often alone on the rig floor while the roughnecks are occupied elsewhere with routine maintenance duties.

Making a connection

When the kelly has drilled down:

1. The driller stops the rotary, raises the drillstring off bottom, and shuts down the mud pump

2. The floorhands set the slips and the driller slacks off (lowers) the drillstring until it is suspended from the slips

3. Using tongs, the floorhands break out the kelly from the drill-string

4. The kelly assembly (which includes the kelly, kelly bushing, swivel, and rotary hose) is swung over to the mousehole, stabbed into the joint of drillpipe standing in the mousehole, and the joint is made up

5. The kelly assembly with the new joint attached is picked up, swung over to the drillstring, stabbed, and made up

6. The string is picked up, the slips pulled, the mud pumps started, the bit run to bottom, and rotation resumed

Roundtripping

It's occasionally necessary to come out of the hole with the drillstring to change the bit, add drill collars, etc. and then run back to bottom. This is called "making a roundtrip," and it proceeds as follows:

Tripping out

1. The slips are set, the kelly is broken out, the assembly is set back in the rathole, and the swivel is unlatched from the hook.

2. The elevators attached to the hook are lowered and latched onto the drillstring below the box that is "looking up".

3. The string is picked up the length of a stand (usually three joints) and the slips reset.

4. The floorhands break out the joint, push the pin end of the stand off to the side, and set it down on wooden sills.

5. The derrickman unlatches the elevators and leans the stand back against the fingerboard. The elevators are run back down

to pull another stand and the process repeated until the entire string is racked in the derrick.

Tripping in. The process is reversed to go back into the hole. The derrick-man latches the elevators on the stands, the stands are picked up, stabbed, and made up into the string.

The handling of each stand is a complex operation involving many sequential steps, and the sequence needs to be repeated over and over. A rhythm develops, requiring split–second timing between the driller, the floor hands, and the derrickman. It's actually safer to work quickly in this smooth, deft, manner with complete concentration than at a slower pace, allowing the mind to wander.

Automated rigs

There are a number of experimental (and a few commercial) applications under way using rigs with robots handling the pipe, thereby largely eliminating floor hands. This technology is driven by the high injury rates inherent in handling tongs, etc. and may some day result in wholesale automation of drilling.

Well Control

Blowouts

Blowouts, where reservoir fluids blow up the hole and out onto the rig floor, are what drillers worry about most. The force of the escaping fluids can smash tools together, causing a spark that could ignite the fluids and burn up everything in the vicinity. Careful attention is therefore given to well control—keeping reservoir fluids out of the wellbore.

The fluids in porous and permeable rocks are under pressure. When these rocks are drilled, this reservoir pressure will force the fluids into the wellbore unless the reservoir pressure is offset by mud pressure. It is essential to prevent any significant entry into the wellbore by reservoir fluids.

Mud pressure reflects the hydrostatic head of the column of mud in the hole above the formation. That is, the downward force of the column of mud. Pressure is expressed in pounds per square inch (psi), which means the weight pressing on one square inch of area.

Mud pressure increases with the depth of the hole and with the density of the mud. For example, the mud pressure at the bottom of a 7000 foot hole filled with mud having a density of 10 pounds per gallon can be calculated as follows (FIGURE 6–11):

1. Convert the gallons to cubic feet.

$$\frac{10\ \text{lbs / gal}}{7.48\ \text{gal / ft}^3} = 74.8\ \text{lbs / ft}^3$$

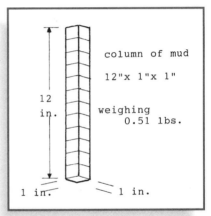

2. Divide by 144 to get the weight of each 12" by 1" by 1" column within the cubic foot. This is the hydrostatic gradient—the increase of mud pressure per foot of depth.

$$74.8 \div 144 = 0.515\ \text{psi (lbs / in}^2)$$

FIG. 6–11 Hydrostatic Gradient

3. Multiply the per–foot hydrostatic gradient by the depth to get the mud pressure at the bottom of the hole.

$$0.515 \times 7000 = 3605\ \text{psi}$$

If the reservoir pressure in this example was 3500 psi, the hole would be overbalanced, meaning that the mud pressure exceeds the reservoir pressure. The pressure differential from the wellbore into the formation would cause some mud filtrate to invade the formation. The solids from the mud filter out on the face of the formation, forming a filter cake. If the reservoir pressure were higher than the mud pressure, say 4000 psi, this under–balance would cause formation fluids to enter the wellbore. If oil or gas enter the wellbore—a kick—aggressive action must be taken quickly to prevent a blowout.

Anatomy of a blowout

1. An underbalanced condition exists, permitting oil or gas to enter the wellbore.

2. The oil and gas displaces mud in the annulus. Since the oil and gas are lighter than the displaced mud, the mud column becomes even lighter. This accelerates the rate of fluid influx.

3. As a gas kick rises up the annulus, the mud column above it becomes shorter, which reduces pressure on the gas. This causes the gas to expand, which displaces more mud and further accelerates the progress of the kick.

4. The kick reaches the surface and blows out through the rig floor. The blowout often ignites, burning the crew and destroying the rig.

Preventing blowouts

Every effort must be made to anticipate the reservoir pressures that will be encountered, and to design the mud program to handle those pressures. This is relatively easy to do in development drilling where the local pressure regime is well known from earlier wells. It is much more difficult when drilling wildcat wells.

It might seem that the best approach to blowout prevention would be to drill the entire hole with mud so heavy that there would be no chance of a kick getting started. This is, however, not practical because heavy mud significantly retards the bit's penetration rate. The practice is therefore to drill the non–permeable sections of the hole somewhat under–balanced, then to weight up for the permeable zones.

The crew must stay alert during drilling to detect the following early signs of a kick developing. These are generally quite subtle and easy to overlook.

- Mud returns cut with oil, gas, or water
- An increased penetration rate caused by underbalance

- A gain in mud pit level from mud displaced by the kick

- A change in the rate of mud return; increasing rate could signal a kick; decreasing rate could signal lost circulation

- A decrease in mud–pump discharge pressure from the lighter mud column

- An increase in drillstring weight; since gas– or oil–cut mud is less dense, it exerts less buoyancy on the drillstring

If a kick is detected early, the mud can be weighted up and the kick circulated out of the hole with little difficulty. When a kick is gaining momentum with the possibility of blowing out, it may be necessary to close the blowout preventer (BOP).

The BOP is a safety valve at the top of the hole, just underneath the rig floor. When activated, the BOP closes and seals off the well, preventing it from blowing out. BOPs are equipped with alternative sets of rams. One set closes around the drillpipe to seal off the annulus. Another set seals off the entire hole when the drillstring is out of the hole. A third set can cut through the drillpipe and seal it off.

Controlling a blowout

The art and science of capping blowouts was significantly advanced by dealing with the hundreds of Kuwaiti wells blown up in the Gulf War.

Capping procedure

1. Working under a cooling spray of water, workers clear away obstructions and debris from around the well before the fire is extinguished. If this isn't done, the fire will re–ignite from the hot metal. There is no urgency to put the fire out because it protects the workers by burning off the oil and gas. This is particularly true when the wellstream includes poison hydrogen sulfide gas.

2. Explosives are suspended over the well with a crane line and detonated. The explosion momentarily consumes all the oxygen, which puts out the fire.

3. If a stub of undamaged pipe remains at the top of the hole, it is prepared to have a valve slipped down over it. This usually involves sawing off the pipe to get below the damage.

4. An open valve equipped with gripping devices is swung over the wellstream with the crane, lowered down over the pipe stub, and secured. The valve is then closed and the well is under control.

5. If there is no pipe available to hold the valve, a relief well must be drilled. A rig is set up off to the side and drills a directional hole that penetrates the formation very near to the original well. Water or cement is then pumped down the relief well to kill the blowout (FIGURE 6–12).

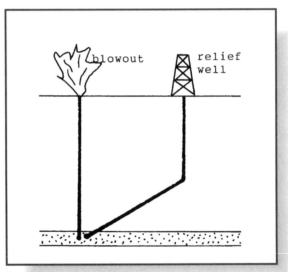

FIG. 6–12   Relief Well

Stuck pipe

Getting the drillstring stuck in the hole is a constant concern for drillers. It happens quite often, almost routinely, but is normally resolved without serious consequences. In some cases, however, a portion of the hole has to be abandoned.

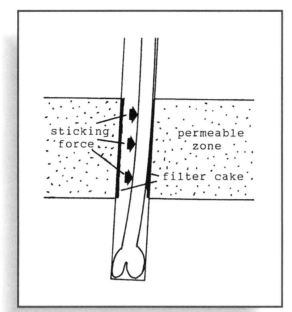

FIG. 6–13 Differential Sticking

Differential sticking. Differential sticking is probably the most common cause of stuck pipe. It is a major problem when drilling softer, unconsolidated formations typical of continental margins (FIGURE 6–13).

When a permeable formation is penetrated while drilling overbalanced, some of the liquid in the mud is forced out into the formation, leaving behind the mud solids as a filtercake on the formation face. Differential sticking occurs if the pipe remains motionless in the hole for a short period of time—such as while making a connection. The pipe rests on the side of the hole, pressing through the filter–cake and contacting the formation face. It becomes differentially stuck when the filter–cake build–up blocks communication of mud–column pressure to the area where the pipe contacts the formation. The pressure in the contact area then drops toward the formation pressure. This creates a pressure differential between the mud column pressure on the inside of the hole and formation pressure on the bottom side. The differential presses the pipe against the side of the hole with such force that it cannot be moved.

Avoiding differential sticking. While overbalanced, the pipe must be kept moving, either reciprocating or rotating, so that differential pressure doesn't build up. It's also helpful to use low filtrate–loss mud to minimize filter–cake thickness.

Hole caving. Several conditions can cause the side of the hole to cave in and, in severe cases, cause the pipe to stick (FIGURE 6–14):

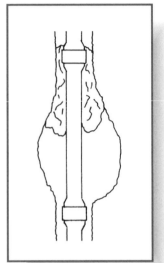

- Some shales absorb mud filtrate, swell up, and slough into the hole.

- Large quantities of material from uncemented or fractured formations can fall into the hole.

- The weight of the overlying rock can extrude salt and plastic shale formations into the wellbore.

FIG. 6–14   Pipe Stuck by Formation Caving

Freeing stuck pipe

1. When differentially stuck, the first approach may be to reduce mud weight and eliminate the overbalance that caused the problem. This is only possible when there is no concern about a permeable zone kicking. If this is unsuccessful, and if mud circulation is still possible, a "pill" of a few barrels of refined oil or mudcake solvent can be circulated down the pipe and up the annulus until its "spotted" over the stuck area. Tension and torque are kept on the pipe for a considerable period to see if it will free up.

2. If still stuck, the next approach may be to run a free–point indicator down the drillstring by wireline. Tension is pulled on the drillstring and the free–point indicator locates the point below which the pipe is not in tension. This is where the pipe is stuck.

   A light explosive is run down the inside of the pipe and detonated immediately above the free–point. The purpose is to "rattle" (loosen) a coupling so that it can be unscrewed. The string is then carefully rotated counterclockwise to break out the rattled coupling. The string is pulled, leaving the fish (bit and lower part of the drillstring) in the hole.

A set of jars (device that delivers sharp blows) is added, the string is re–run and stabbed back into the fish. The jars are activated, perhaps repeatedly, to deliver upward hammerblows to the fish. In most cases, it is quickly dislodged and recovered.

3. If jarring fails, a wash pipe may be run to wash over the outside of the fish, removing the material that is causing it to stick.

Twist–offs

Drillpipe failures, called twist–offs, sometimes occur from metal fatigue during routine drilling operations. Fishing operations are then required as follows:

1. The condition of the portion of the fish that is looking up must be determined before a strategy for latching on to it can be devised. The piece of parted pipe that was recovered provides a great deal of information about the condition of the piece looking up, but sometimes it is necessary to run an impression block. This has a flat bottom of soft metal that is bumped down on top of the fish. The fish indents the soft metal, providing a picture of its degree of distortion and raggedness. The impression block also shows whether the fish is standing up straight in the hole or is leaning over against the side.

2. It may be necessary to dress–off (clean up) the fish so that the fishing tools can get over or into it. This involves running mills with tungsten carbide surfaces for grinding steel. If an overshot (fishing tool that fits over, or swallows, the fish) is to be used, a mill that grinds down over the outside of the fish is run. If a spear (fishing tool that fits down inside the fish) is to be run, a tapered mill is used to penetrate inside the fish (FIGURE 6–15).

3. When the fish has been dressed, the fishing tools and jars are run on the drillstring. If the fish is standing up, clear of the

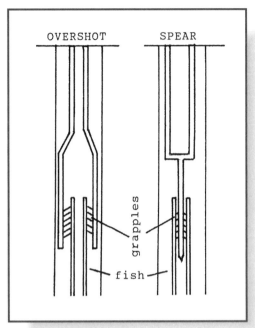

FIG. 6–15   Fishing Tools

side of the hole, overshots can be run. They are preferred to spears because it's difficult to unlatch spears if the fish can't be moved. This can result in a bigger fish.

4. The fishing tools, either overshots or spears, have multiple grapples that grip the fish securely. The jars are then activated and hopefully dislodge the fish.

Directional Drilling

Directional drilling was developed when the industry moved offshore several decades ago. Whereas an onshore rig can easily be moved to a new location to drill another vertical well, in offshore waters the horizontal space needed to support the rig must be created by erecting a platform. In deep waters, the cost of a platform is so great that it is not feasible to build one for each well. Directional drilling technology was therefore developed to drill multiple wells from the same platform, but with the wells having dispersed bottom–hole locations .

Directional wells drilled from off-shore platforms typically have S–shaped profiles. They start out vertically, then kick off at an angle until the lateral displacement (called reach) is attained, then return to vertical until the reservoir is penetrated (FIGURE 6–16).

Bent–sub assembly

The bottom–hole assembly (BHA) used for directional drilling incorporates a downhole mud motor with a bent sub that forces the bit to the side of the hole, thus building angle (FIGURE 6–17). When

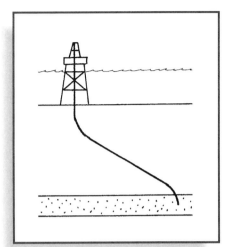

FIG. 6–16   Directional Drilling

the motor operates, it rotates the bit while the drillstring remains stationary. This permits directional control. Using a magnetic insert in the bent sub, the BHA can be precisely oriented with respect to azimuth (North, East, etc.). The inclination of the borehole is also surveyed regularly. With these two pieces of information, the bottom–hole location is continuously calculated and the well can be accurately steered to its target. Devices can also be placed in the BHA, immediately above the bit, to sense azimuth and inclination, and continuously transmit the information to the surface via pulses in the mud column.

Since more horsepower can normally be brought to bear on the bit by conventional drillstring rotation than by using the mud motor, the initial vertical section of hole is usually drilled conventionally, with the mud motor installed when kick–off depth is reached. Once the required kick–off angle has been built with the motor, the straight, diagonal leg of the hole is drilled using both the motor and drillstring rotation. The motor is then used to build angle back to the vertical, and drillstring rotation is resumed to finish the final vertical section.

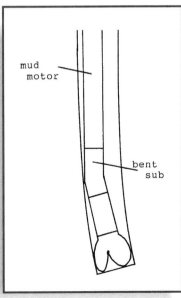

FIG. 6–17   Bent-Sub Assembly

Horizontal drilling

Directional wells are sometimes drilled to penetrate the reservoir at a very shallow angle or are drilled horizontally for some distance within the reservoir. To date, the usual reason to drill horizontally is to expose more of a low–permeability reservoir to the wellbore so that production rate is increased. Any low–permeability reservoir is a potential target for horizontal drilling, but it works particularly well in low–permeability reservoirs with vertical fractures that can be connected by the horizontal borehole. This is the case with the Austin Chalk development in Texas—the most intensively horizontal drilled reservoir in the world.

Horizontal drilling is developing rapidly and the cost is coming down, but it still is considerably more expensive than conventional drilling. Horizontal wells are therefore a small fraction of the total wells drilled, with the technology reserved for special cases such as:

- In the North Sea, the large, expensive platforms that were set to develop giant fields are now being used to drill horizontal wells to develop small fields in the vicinity that would otherwise be uneconomic.

- Reservoirs with thin oil columns are often troubled with bottom water "coning" upward and covering the perforations so that only water is produced. Horizontal wells, by distributing the pressure drop across a broad area, have been found effective in preventing coning.

The tools and procedures used in steering the bit along a horizontal trajectory are not greatly different from those used in conventional directional drilling. Bent subs with downhole motors are used with drill collars for stiffness and weight and with stabilizers (large diameter drill collars that contact the side of the hole) to act as pivot points (FIGURE 6–18).

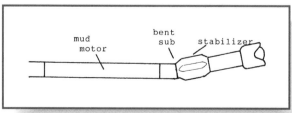

FIG. 6–18 Stabilizer Assembly for Horizontal Drilling

Laterals. Horizontal drilling technology is increasingly being used to drill short lateral extensions from the same borehole (FIGURE 6–19). Some of the applications for laterals are

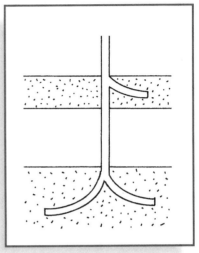

- Replacing an original completion that has been damaged

- Increasing a well's production rate from a tight formation by using opposed multiple laterals to achieve greater access to the formation

- Completing a well in multiple zones using stacked multiple laterals

FIG. 6–19 Multiple Laterals

Drilling Rig Conveyances

The drilling rig itself is essentially the same anywhere in the world, whether onshore or offshore. It's the delivery system, which moves and supports the rig while drilling, that changes greatly with the circumstances.

Land rigs

Jackknife rigs. The jackknife rig is used to drill most land wells. To move the rig, the mast is lowered (jackknifed) and the entire rig is broken down into pieces small enough for transport by truck.

The drilling floor is usually elevated, with the BOP stack fitting between the ground and the floor. The practice of digging a cellar for the BOPs with the rig floor at ground level has largely been discontinued. It was expensive and the cellar could collect poisonous hydrogen sulfide gas.

If the drilling location is on dry ground, its preparation often involves only leveling it with a bulldozer and lining earthen pits with plastic. In some cases gravel, shell or other hard material needs to be added. If the location is wet and likely to become muddy, it may be necessary to cover it with wooden boards.

Helicopter rigs. When drilling wildcat wells in remote areas, there often are not adequate roads available to truck in the rig. Special rigs are therefore manufactured for this situation that break down in pieces small enough for helicopters to sling–lift them onto location.

Offshore rigs

Barge–mounted rig. In shallow bays and canals dredged through marshes, the rigs are mounted on steel barges that are pushed into place and flooded so that they set on the bottom. When the well is completed, the water is pumped out to float the barge and tow it to the next location (FIGURE 6–20).

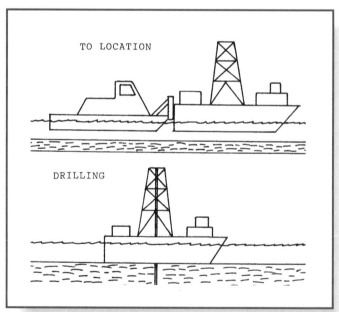

FIG. 6–20 Barge-Mounted Rig

Jackup rigs. Jackups are highly mobile, bottom–supported rigs used extensively in water depths up to 400'. They have seaworthy hulls and are towed to location with the legs (usually three) jacked up into the air. On location, the legs are jacked down to the sea bottom and the hull then jacked up out

of the water. The air–gap between the hull and the water must be adequate
to prevent waves from hitting the hull (FIGURE 6–21).

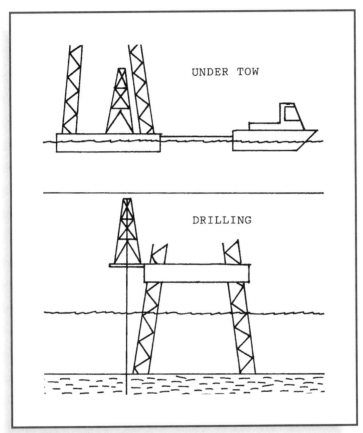

UNDER TOW

DRILLING

FIG. 6–21 Jackup Rig

Jackups are used to drill both exploratory and development wells.
Development wells are usually drilled by jacking up adjacent to a pre–set
well platform, cantilevering the rig out over the platform, then drilling down
through it.

Tender–supported rigs. When permanent production platforms are set in
relatively shallow waters with mild sea states, development drilling is often
done with temporary platform rigs supported by tenders. Tenders are
ship–shape vessels that are towed to location and anchored beside the platform (FIGURE 6–22).

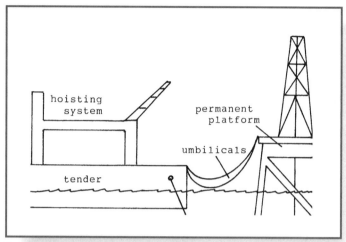

FIG. 6–22 Tender-Supported Rig

Using the tender's crane, the mast and drawworks are lifted from the tender onto the platform. The tender holds the mud pits and pumps, cementing tanks & pumps, mud & pipe storage, living quarters, prime mover, etc. Multiple hoses and power lines deliver the tender's resources to the rig.

After the platform is drilled up, the rig is lifted back on the tender and the tender is towed to a new platform.

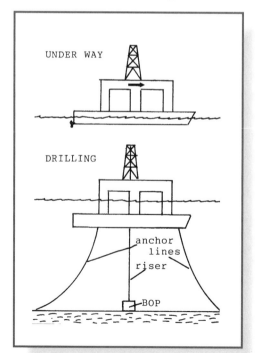

Semi–submersible rigs. Semis are floaters (not bottom–supported) used to drill in deeper offshore waters. They are self–propelled and entirely seaworthy, so they can operate anywhere in the world. Semis normally are tethered over the location by multiple anchors, but some are equipped with thrusters on their four corners so that they can be dynamically positioned. Because of their large size, semis are very stable in high seas (FIGURE 6–23).

When drilling from floaters, the BOP is located on the sea floor

FIG. 6–23 Semi-submersible Rig

and a riser pipe connects the well to the surface. The riser is supported by steel cables from the rig. A tensioner system plays–out and reels–in the cable as the rig moves up and down with wave action. The rig also has heave compensators mounted on the traveling block. These keep the bit on bottom, drilling, while the vessel is heaving up and down.

Drillships. Drillships are dynamically positioned, so they are not limited to water depths where anchoring is feasible. The ship–shaped hull permits faster movement from location to location. The disadvantage of the ship–shape is its lack of stability in high seas. Drillships are therefore used extensively to drill wildcat wells around the world in very deep water and for development drilling (FIGURE 6–24).

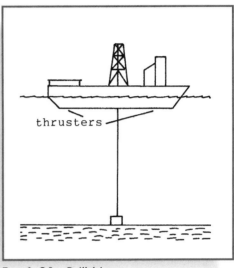

FIG. 6–24 Drillship

Permanent platform rigs. In the North Sea and other areas with demanding sea states, it's often not feasible to lift rigs on and off the platforms. Instead, the rigs are installed onshore before float–out and stay on the platform permanently. In the field's later stages, the rigs are used for workovers.

Contracts and Crews

Contracts

Virtually all drilling rigs are owned by drilling contractors—not by the operating companies. This is because a contractor can move the rigs around the world to keep them busy, whereas the operator shuts down drilling at times and would have to idle the rigs.

Many rig lease contracts, particularly onshore, are turn–key, where the contractor drills a hole of specified size to a specified depth for a given sum of money. The price may be expressed in $/ft. Day–rate contracts, particularly prevalent offshore and in international operations, are expressed in $/day for the lease of the rig.

Crews

The rig's operating crew are employees of the drilling contractor. They work 8– or 12–hour tours (pronounced "towers"). The rig operates 24 hours a day, 7 days a week.

Operating Crew

- The toolpusher is in charge of the overall operation.

- The driller is the foreman of the drilling crew and also the hands–on operator of the rig.

- The drilling crew are called roughnecks. During a roundtrip, one roughneck functions as the derrickman, while three or four function as floor hands; between roundtrips, they have varied duties checking and maintaining all the rig equipment.

- Some rigs, particularly offshore, employ roustabouts for general labor; the head roustabout is typically the crane operator, who reports to the toolpusher.

- Big rigs may employ specialists such as motormen, mechanics, and electricians.

Support Personnel

- The company man represents the operating company that hired the drilling contractor.

- The drilling engineer employed by the operating company prepares the well plan and provides engineering input as needed during operations.

- The mud specialist monitors the condition of the mud and recommends changes to it.

- The wellsite geologist keeps track of what rock formations the bit is penetrating.

- The mud logger monitors shows of oil and gas in the returning mud stream.

FORMATION EVALUATION 7

Data Requirements

Drilling a well not only provides an opportunity to produce oil or gas, it is also an opportunity to collect information about the formations being penetrated. In fact, expendable holes are often drilled solely to gather information with no intent to ever complete them.

The data is gathered at different stages. Data from cores, drill–stem tests, and mud logs is obtained while the hole is actually being drilled. Open–hole well logs are run after total depth is reached but before the casing is run. Reservoir sampling takes place after the well is completed and equipped, but before it is produced. The 24–hour initial potential–test at the start of production is the final piece of data from a new well.

Since there are costs associated with each type of data gathered, not all the possible data is gathered from each well. A new–field wildcat, for example, justifies gathering a great deal more data than does an in–fill development well in a mature field.

Wildcat wells

When drilling wildcat wells, the pressing need is to recognize potentially productive zones as they are being penetrated. Because the weight of the mud column prevents reservoir fluids from entering the wellbore, an oil or gas zone may be drilled through unnoticed unless cuttings and return–mud data are carefully gathered and analyzed. Most oil companies have had the experience of abandoning a supposedly dry wildcat, releasing the acreage, and then having a competitor pick up the acreage and drill a major discovery.

Development wells

Information gathered from each additional development well helps assess the reservoir's overall extent and quality. Of particular interest are porosity, permeability, fluid saturations, pressure, zonal thickness, lithology, and production rate.

Drilling Operations Log

The drilling operations log is a continuous, foot–by–foot record of the hole as it is being drilled. All pertinent data gathered from the cuttings, mud returns, and drilling operations are plotted against depth on the same strip–chart so that they can be interpreted simultaneously. The log is maintained by the operating company's wellsite geologist, if one has been assigned. Otherwise, the contract mud logger does it.

The operations log helps the wellsite geologist track the well's progress down through the stratigraphic column. The process is much like that of a navigator continuously plotting an airplane's position. It's essential that the geologist anticipates when the bit is approaching each zone of interest so that it can be thoroughly evaluated. This is particularly demanding with rank wildcats where little is known about the stratigraphy.

Cuttings data

When drilling consolidated, harder, and older rocks typical of inland locations, analysis of the drill cuttings provides valuable information about the formations being penetrated. The cuttings from unconsolidated, softer, and younger sediments typical of coastal and offshore areas, however, tend to disintegrate and provide less information.

The wellsite geologist monitors the bit's progress by examining the cuttings that rise to the surface in the mudstream and are filtered out on the shale shaker. A sample of the cuttings might be collected every ten or fifteen feet during most of the drilling, but samples may be taken every foot when close to a pay zone.

The geologist washes the samples, then examines them under a microscope. To determine exactly where the bit is in the stratigraphic column, the geologist tries to spot the very first cuttings from each new zone drilled. This is a difficult task where experience counts greatly. Some of the complications are

- A significant lag time, often several hours, exists between the cuttings being cut and their reaching the surface. The geologist must correct for the time lag before posting a depth on the sample log.

- Cuttings of differing sizes and densities rise up the annulus at different rates. Large, low density cuttings rise swiftly, while small, high density cuttings tend to rise slower or even to fall back. A given sample may include cuttings that were cut both a week ago and an hour ago. Particles also slough off from formations exposed up the hole. The geologist must be able to recognize and discard material that he's seen previously, focusing only on the new cuttings.

- If the mud's gel strength is inadequate to keep the hole clean, cuttings may fall back under the bit and be re–drilled to such a small size that they cannot be analyzed.

Lithology. The cuttings are examined under a microscope and the lithology is described in detail. For example, a cutting might be described as "brown, poorly–sorted, angular, medium–grained sandstone with light siliceous cementation." From this description, the geologist can find the equivalent zone in nearby wells to locate himself in the stratigraphic column. This is called "correlation." Note that much can change between wells as some sections thicken, others thin, and structural displacement occurs by folding or faulting. The sample logs are key to correlating zones.

Porosity. Porosity visible under the microscope is noted on the sample log.

Oil shows. Although the cuttings are thoroughly flushed by mud filtrate during drilling, traces of an oil saturation are usually retained in the tight corners of the pores. This can appear as a stain under the microscope, or show up as colored fluorescence when the porosity is exposed to ultraviolet light. The geologist notes the percent coverage and intensity of stains and fluorescence. To further analyze shows, the sample may be ground up, cut with a solvent, and the color of the resulting liquid noted. The color of a cut and its fluorescence gives an indication of the gravity of the oil, with heavier oil having a darker color.

Gas shows. The cuttings can be analyzed for gas, called microgas, by pulverizing them and running the released vapors through a gas analyzer.

Hydrocarbon odor. When the cutting's sample container is first opened, the geologist sniffs it to detect any hydrocarbon odors.

Drilling data

The drilling–time log, in minutes per foot, is plotted on the left side of the operations log. Its function is to signal increases in the bit's penetration rate, called drilling breaks. A sharp break can indicate the bit has drilled out of non–porous rock into porous rock.

The value of drilling time is that it gives a real–time indication of porosity. In contrast, by the time the cuttings from a new zone get to the surface and are analyzed, the zone may already have been penetrated, precluding cutting a core or running a drill–stem test. When a drilling break occurs, the bit can be picked up and the hole circulated "bottoms–up." The last cuttings can then be analyzed and the decision made whether to core or test.

The drilling crew also notes any gas breaking out of the mud when it reaches the surface, or oil in the mud or floating on the mud pits. These shows can be valuable indicators that a hydrocarbon reservoir is being penetrated.

Mud data

Gas shows. Mud–logging units continuously monitor the returning mud stream for traces of natural gas. This gas is not from a kick—gas entering the wellbore from the formation. Instead, it is the tiny amount of gas in the pores of the rock drilled as the bit penetrated the reservoir. In the reservoir it may have existed either as free gas or have been dissolved in the oil.

Oil shows. Oil traces in the returning mud stream can be detected by exposing mud samples to ultraviolet light and noting any fluorescence. With oil–base mud, however, fluorescence from its refined oils may interfere with detecting shows of light crude oils.

Coring

Coring operations capture a sample of the actual subsurface rocks. Other formation–evaluation techniques are somewhat indirect, but cores are very direct. They can be picked up, examined in detail, smelled, weighed, and analyzed for key reservoir parameters. A core has lasting value, often being subjected to additional analysis late in a field's life when secondary or tertiary reservoir–drive projects are being considered.

Conventional coring

In conventional coring, the drillstring is pulled out of the hole and the drillbit is replaced with a coring assembly. This consists of a doughnut–shaped (circular with a hole in the middle) diamond or PDC bit run on a hollow core–barrel. The string is run back to bottom, rotation and mud circulation started, and drilling resumed (FIGURE 7–1).

As the bit penetrates, a solid core of undrilled rock rises up through its center into the barrel. When the zone of interest has been penetrated or when the core barrel is full (barrels are usually 30' long, but can be longer), the string in pulled. A spring–loaded core–catcher in the barrel grips the core so that it stays in the barrel and is recovered at the surface.

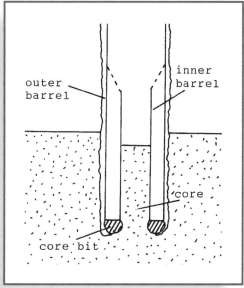

FIG. 7–1   Conventional Core Assembly

Cutting a core requires a round–trip to install the coring assembly and a second round–trip to remove it. Round trips occupy a great deal of rig time, particularly in deep wells. Additional time is lost because a core bit drills much slower than a conventional bit. Coring is therefore very costly, which results in its being used sparingly.

Conventional coring works only in the consolidated, competent rocks typical of older, onshore formations—so–called "hard rocks." The unconsolidated formations typically found along continental margins are too friable to be recovered by conventional coring methods.

When to start coring. Wildcat well–plans in hard–rock country frequently specify that all porosity breaks encountered in the target formations be cored.

The wellsite geologist doesn't know if or when any porous intervals will be encountered. He therefore carefully monitors the progress of the bit and shuts down the drilling immediately at the first indication of porosity. If he is not alert, the zone will be drilled through and the opportunity to cut a core lost forever.

A break on the drilling–time log is often the first indication of porosity. When the geologist sees a break, he has the driller stop rotating, pick the bit up off bottom and circulate bottoms–up. This lets him examine the last cuttings drilled to see if a porous section has indeed been encountered, and if he can detect any shows of hydrocarbons. If it looks good, he has the driller trip out to pick up the coring assembly.

There is a great deal of pressure on the geologist, first, to not miss a potential pay zone and, second, to not waste rig time in coring shales or other non–reservoir rocks. When a core–barrel full of shale is recovered, the geologist's credibility suffers and he may experience resistance from the drillers the next time he wants to cut a core. Since the drilling crew's primary interest is in drilling the hole as quickly as possible, they can be impatient with the geologist's using rig time for gathering data.

Core analysis. The cores are taken to a core laboratory for analysis. Whole–core analysis is sometimes performed, but plug analysis of smaller samples cut out of the whole core is more common. The following laboratory analyses are routinely run on the samples

- *Porosity* is determined by several methods; one method uses a retort to boil off the saturating liquids, then condenses the liquids and compares their volume to the plug volume.

- *Fluid saturations* can also be obtained from the condensed retort fluids. Because the core was flushed by mud filtrate during drilling, the S_0 cannot be determined quantitatively— only qualitatively.

- *Permeability* is determined by sealing the plug sample in a pressure housing and forcing fluid through it.

Both vertical and horizontal permeability are usually measured since vertical permeability, particularly in sandstones, is often less than horizontal permeability.

Sidewall coring

Sidewall coring guns are run on wireline. The cores are taken from the side of the open hole by explosively shooting small cylinders into the formation, capturing formation material in the cylinders. The cylinders are on wire tethers and are recovered with the gun when it is pulled (FIGURE 7–2).

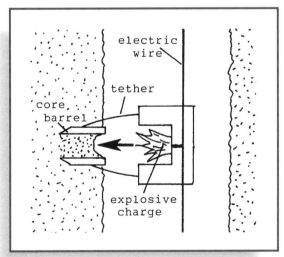

FIG. 7–2   Sidewall Coring with Wireline

Sidewall cores are obtained only from relatively soft, unconsolidated formations—the cylinders cannot penetrate hard rock. Since the cylinder's impact causes crushing and compaction of the core, porosity and permeability data are not dependable.

A major use of sidewall cores is to supplement the open–hole logs by checking specific spots where the log interpretation is unclear regarding lithology or fluid saturations.

Productivity Tests

Obtaining an actual flow test on a potential reservoir has unique significance. Although the well logs might indicate that a reservoir has the necessary permeability and saturations to produce commercially, the logs are only indirect indications and can be misleading. Until oil or gas is actually produced, an operator can never be completely certain of a well's productivity.

Some very costly development mistakes have been made using well log data only. The productivity of gas reservoirs is particularly difficult to assess without extended production testing. Productivity testing is so important that the Society of Petroleum Engineers (SPE) requires that the commercial producibility of reserves designated as proved be supported by actual production or formation tests.

Drillstem testing

Drillstem tests (DSTs) are used to evaluate possible producing zones encountered in wildcat wells. Both fluid–flow and reservoir pressure data is provided by DSTs. The normal sequence of events is

1. The drillstring is pulled to replace the drillbit with the test assembly. The assembly includes valving, a packer to seal off between the drillpipe and the open hole, and a pressure recorder located below the packer (FIGURE 7–3).

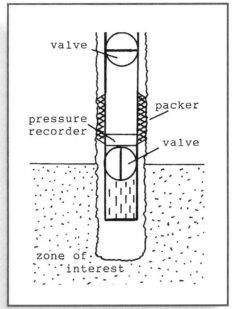

FIG. 7–3   Drillstem Test Tools

2. The test assembly is run on the drillstring with the tool closed so that no mud enters the drillstring. The packer is set just above the zone of interest.

    While going in the hole, the pressure recorder picks up the increasing weight of the mud column. When the packer is set, the mud weight is removed from the tool.

3. The valving is then shifted so that the pressure recorder is exposed to the reservoir for about 15 minutes to record the initial reservoir pressure.

4. The valving is again shifted so that any fluids the reservoir will produce flow up the drillstring. This flow period usually lasts half an hour or longer and is a period of high drama. With luck, gas or oil will flow to the surface, establishing a discovery.

    On the rig floor, everyone gathers around the open end of the drill pipe watching for signs that fluid is entering the pipe. The first sign is a blow of air which indicates that fluid—air, water, or gas—is displacing the air in the pipe. This establishes that the zone has permeability—an encouraging sign.

The attention then focuses on determining which fluid is being produced.

A rapidly increasing blow of air is an indication that the zone is producing gas. If so, gas will reach the surface in a few minutes and can become a hazard. The drillstring is therefore quickly hooked up to a manifold which diverts the gas flow away from the rig floor and through a pipeline to a downwind flare.

If the blow of air maintains a relatively steady flow, it's likely that the zone is producing oil. The oil may or may not reach the surface, depending on the reservoir pressure. Oil flowing to the surface is generally the best possible outcome of a flow test.

If the air blow slows and stops before fluids reach the surface, it's likely that saltwater, or a mixture of saltwater and oil, is being produced. Because of its greater density, saltwater rarely flows to the surface.

If oil or water is produced during the flow test, the recorded pressure shows the buildup of hydrostatic pressure from the rising fluid level in the drillstring.

5. At the end of the flow test, the valving is again shifted to obtain a final reservoir pressure.

    The rate of the pressure buildup is more important than the pressure itself. A rapid buildup shows that influx from the reservoir quickly replaces the fluids produced during the flow test. This indicates high permeability—a major criteria of reservoir quality.

6. The packer is then released, and the drillstring pulled. The string is pulled wet, so that as each stand is broken out the liquids in it gush out over the drill floor. The number of stands that contain saltwater and oil help evaluate the formation.

    When the test assembly reaches the surface, the chart from the pressure recorder is recovered and the pressure information read from the chart (FIGURE 7–4). New technology is available which instantly transmits the pressure data to the surface via acoustic pulses up the mud column. This is an expensive service, but allows the tester to vary the length of the flow and final pressure periods in response to the observed reservoir performance.

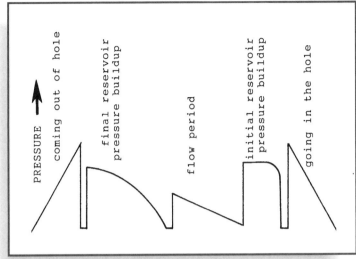

FIG. 7–4   Drillstem Test Pressure Chart

It is often difficult to get a good packer setting because of the irregularity of the sides of the hole. Many DSTs fail because the packer doesn't hold. This problem is greatly magnified when the formation to be tested has been drilled completely through. Tandem packers are then required to isolate the formation—one set above the and one set below. The likelihood of getting tandem packers to hold is minimal. As with coring, it's essential that the wellsite geologist stop the bit's penetration before the zone is completely penetrated.

DSTs are also run in holes that have been drilled to total depth (TD), cased, cemented, and then perforated. This avoids packer–seating problems and the possibility of getting stuck that is always present in open–hole wells. It also provides higher–quality test data. In some situations, these advantages may justify the added costs of casing, cementing, and perforating.

Wireline formation testing

Wireline formation testing is a far less expensive method than DSTs to obtain fluid–flow and reservoir pressure data, but the data is of considerable less quality.

After TD is reached, but prior to running casing, the test tool is run on an electrical–conducting wireline. A spring forces the tool into contact with the side of the hole so that formation pressures and flow samples can be taken. Multiple zones can be tested on a single run into the hole.

Initial potential (IP) test

When each well is completed, equipped, and all its load fluids (fluids pumped into the well during stimulation, killing, etc.) are recovered, an IP test is run to determine its initial productivity.

The test period is usually 24 hours, but shorter tests are often taken and erected to a 24–hour basis. The produced barrels of oil, barrels of water and MCF of gas (or metric equivalents) are reported along with the tubing pressure and choke size if the well is flowing. If the well is artificially lifted, the specifications of the lift equipment are included.

A great deal of attention is paid to the IP test since it often is the first dependable indication of a well's productivity, and productivity largely determines profitability.

The IP is usually the signal to remove the well from the drill report. It marks the well's change from drilling status to production or other status.

Open–Hole Well Logs

Multiple open–hole logs are run on all wells, including those that have been cored or drillstem tested. Log data constitutes the overwhelming bulk of formation evaluation data that is gathered on most wells.

Logging uses some of the most advanced technology in the upstream industry (exceeded perhaps only by seismic technology). There is constant development of new tools, improvement of old tools, and refinement of interpretation techniques.

Each of the various logging tools measures one or more of the electrical, acoustic, or radioactive characteristics of the formations or formation fluids. Some of the signals captured are emitted spontaneously by the formation and some are induced by emissions from the logging tool.

Wells are normally logged in the open hole before casing is run because the pipe interferes with most log measurements. Some radioactive tools, however, can operate in cased holes.

Most routine log interpretations are quite simple and can be performed by someone with minimal training, but professional log analysts, also called "petrophysicists" are required for the more demanding interpretations.

Logging operations

The logging sondes are run into the hole on an electrical–conducting wireline spooled off of a drum mounted in the back of a logging truck (onshore) or mounted on a skid (offshore).

Several sondes are typically combined into a single assembly and run together. Multiple runs are required to obtain all the logs required on particularly significant wells. Logging can be a major part of a well's cost (FIGURE 7–5).

The sonde is run to the bottom of the hole, then slowly retrieved as the actual logging takes place. The logged interval extends from the bottom of the hole up through the shallowest zone of interest; the rest of the hole is not logged. For example, logs on a 10,000 ft. hole might cover only the bottom 2,000 ft. The log measurements are normally printed out on an 8–inch strip chart that is folded in accordion pleats.

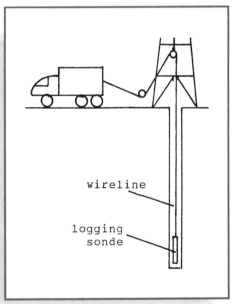

FIG. 7–5   Well Logging

Lithology (permeability) logs

Spontaneous Potential (SP) log. The spontaneous potential is a naturally–occurring, electrochemical potential present in the borehole at the boundaries between porous, permeable zones and surrounding shales. The SP log is therefore used to identify potential sandstone reservoirs and to establish their thickness and relative quality.

The SP log is plotted on the left of the strip chart, with a leftward inflection of the trace indicating a porous and permeable zone (FIGURE 7–6).

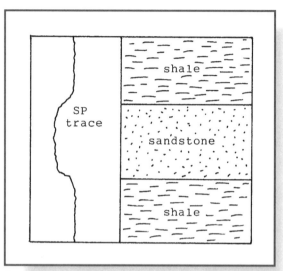

FIG. 7–6   Spontaneous Potential (SP) Log

Gamma ray log. The gamma ray tool contains a scintillation counter that records the intensity of naturally–occurring gamma ray emissions from the surveyed formations. Since shales, particularly marine shales, emit considerably more gamma rays than do other sedimentary rocks, the gamma–ray log is used similarly to the SP log to differentiate sandstones from surrounding shales. The gamma ray trace is positioned on the strip chart similarly to the SP trace, with a leftward inflection indicating porous and permeable zones.

Porosity logs

Sonic (acoustic) log. The sonic logging tool generates an acoustic pulse and measures its transit time through the formation to receivers mounted in the tool some distance below the generator.

The slower the transit velocity, the lower the porosity. Knowing the lithology (sandstone, limestone, etc.), the porosity can be accurately calculated from the transit time.

Density log. The density tool emits gamma radiation into the formation and then records how much of it reaches detectors located several feet above the radiation source. The more gamma–rays that are absorbed in transit through the formation, the greater the formation's density. High porosity reduces rock density, so the more rays reaching the detectors, the higher the zone's porosity.

Neutron log. The neutron tool emits neutrons and measures the rate of their capture by the formations logged. Since it is the hydrogen in water, oil, and gas that captures neutrons, porosity is related to the rate of capture.

Since gas contains less hydrogen than water or oil, the porosity of gas–filled zones reads erroneously low on neutron logs. This effect is often used to differentiate between gas and oil saturations by running both a neutron and a density log and comparing the two porosities. Permeable, resistive zones with neutron porosity substantially lower than density porosity are interpreted as gas saturated rather than oil saturated.

Resistivity logs

The physical principle underlying resistivity logs is that saltwater–saturated porous zones conduct electrical current better than do similar zones saturated with oil, natural gas, or fresh water. This is because saltwater is an electrolyte, having free ions of sodium and chloride that cause the water to be electrically conductive. Since crude oil, natural gas, and fresh water lack free ions, they are more resistive to the flow of electrical current. Resistivity logs are therefore used to differentiate hydrocarbon–saturated permeable zones encountered in a well from the much more numerous saltwater–bearing permeable zones.

Resistivity tools impress an electrical potential across the formation and measure the current flowing through it. The formation's resistivity is then calculated using the relationship

$$\text{resistivity} = \frac{\text{potential}}{\text{current}}$$

On the strip log, the resistivity traces are positioned to the right of the SP or gamma ray trace. Rightward inflection indicates higher resistivity.

Figure 7–7 illustrates how the SP log measures the thickness of the reservoir while the resistivity log indicates the fluid saturations. In this case, the upper part of the sandstone is saturated with oil (highly resistive on the log) while the lower part is saturated with saltwater (less resistive on the log). An oil/water contact is therefore present.

Using porosity from a porosity log and estimates of the resistivity of formation oil, water, and gas, the formation's water saturation (S_w) can be calculated from resistivity logs. Once S_w is known, the all–important hydrocarbon saturations are calculated $(1 - S_w)$ and the interpreter can predict whether the zone will produce oil, water, gas, or a combination. This is the main function of well–logging.

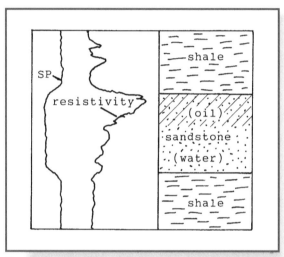

FIG. 7–7 SP and Resistivity Logs Showing an Oil/ Water Contact in a Sandstone Reservoir

Note that the log in Figure 7–7 could as well be interpreted as having gas above the saltwater rather than oil. Both oil and gas are resistive and show up similarly on most resistivity logs. By being familiar with the area, the interpreter probably knows whether to expect gas, oil, or both.

Several different resistivity logs are run in the same well to aid in interpretation. For example, logs with close electrode spacing (i.e., 1.5 inches) are used to investigate the formation near the wellbore, while logs with wider spacing investigate a wider radius. This is useful because near–wellbore porosity is more thoroughly flushed by mud filtrate, which has different resistivity than the formation fluids.

Caliper log

The diameter of the hole can enlarge from the sloughing of soft shales during drilling or be restricted by the buildup of mud filtercake across permeable zones. Competent rocks like limestone or well–cemented sandstones are, however, normally "in gauge."

Several spring–loaded arms on the caliper move in and out as the tool is drawn up the hole, calculating the dimensions and volume of the shaft. The primary uses of the caliper log are to

1. Obtain an accurate hole diameter needed to interpret some of the other logs

2. Identify filter cake that can indicate the presence of permeable zones

3. Determine the hole volume to know the quantity of cement needed to cement the casing

4. Locate good packer seats for drill–stem tests

Reservoir Fluid Samples

As discussed in Chapter 2, an oil reservoir is an extremely complex system that behaves very differently depending on its unique blend of hydrocarbon molecules. When a significant new field is discovered, therefore, a sample of its oil under initial reservoir conditions is often obtained and subjected to P–V–T (pressure–volume–temperature) analysis in the laboratory. The reservoir fluid's unique characteristics are then used to forecast production rate and reserves.

The sample is taken after the well is completed and equipped, but before it has produced significantly—initial pressure conditions are essential to getting a representative sample. The well is first conditioned by producing it at a steady state. It is then shut in, the sampler is run on a wireline under pressure, and a sample is taken at reservoir depth. An alternative procedure takes the sample at the surface while the well is producing.

WELL COMPLETION 8

When a drilling well fails to find any zones of interest, it is declared a dry hole and is permanently plugged and abandoned (P&A) by placing several cement plugs in the hole. Each plug might be 200–300 feet long with drilling mud occupying the space between the plugs.

If a zone showing commercial potential is encountered, the decision may be made to complete the well as a producer. The sequence of events is as follows:

1. If it is not already in the hole, the final string of casing is run and cemented (note that one or more shorter strings of casing have already been run and cemented by the time a drilling well reaches total depth)

2. The casing is perforated opposite the zone(s) of interest

3. If necessary, the zone is stimulated to attain an adequate production rate or gravel–packed to stop sand production

4. The well is equipped with tubing, packers, and a Christmas tree; it is then ready for production

Casing

Casing strings consist of multiple joints of large–diameter steel pipe that are screwed together, one joint at a time, as the pipe is run in the hole.

Unlike tubing strings, which are retrievable, casing strings are permanently set in oil and gas wells with the annular space between the casing and the hole filled with cement.

Purpose of casing

The casing and its cement sheath have several important functions:

1. Casing protects the hole from the mud.

    - Some shales tend to draw water out of the mud, causing them to swell and slough off into the hole.

    - Loose surface sediments and unconsolidated formations are eroded by the mud stream.

2. Surface casing in onshore wells protects near–surface potable water zones from contamination by the deeper saltwater zones; this is not an issue offshore.

3. Casing provides a smooth conduit for running in and out of the hole with tools.

4. Casing and its surrounding cement sheath isolates downhole zones so that they can be produced separately.

    - For optimal recovery from multiple reservoirs in the same hole, each reservoir must be pressure–isolated from the others.

    - Within a single reservoir, oil often needs to be produced from the oil leg while leaving gascap gas or bottom water untapped.

Casing program onshore

As the well is drilled deeper, new casing strings are run and cemented, as required, to protect the hole from further exposure to the circulating mud. The number of strings needed is determined by the depth of the well, the relative stability of the formations being penetrated, and the characteristics of the drilling fluid. Each successive casing string is run inside the previous string, so the diameter of each new string decreases.

Surface casing. The drilling operation begins by drilling the large–diameter surface hole to penetrate unconsolidated surface material and any potable water zones. It is normally a few hundred feet deep. Large–diameter casing is run to the bottom of the hole and cement slurry is pumped down the inside of the pipe and circulated up the "backside" annular space between the casing and hole. Several hours are allowed for the cement to harden.

Intermediate casing. Many wells, particularly shallow ones, do not require an intermediate string, but if there is a formation down the hole that becomes unstable in prolonged contact with the mud, an additional string is necessary. Some deep holes require two or three intermediate strings.

To drill out, a smaller bit that fits inside the surface pipe is run, the drillable shoe on the surface string and cement in the bottom of the hole is readily penetrated, and drilling proceeds. Once the sloughing zone is penetrated, the drillstring is pulled and the intermediate string is run to bottom and cemented. It is important that the cement rises high enough in the annulus to tie into the cement in the surface pipe. This provides an unbroken cement sheath covering the entire length of the hole. Each successive string of casing is hung from the wellhead at the surface. The wellhead also seals off the annulus (FIGURE 8–1).

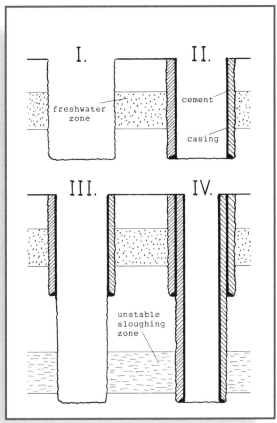

FIG. 8–1   Surface and Intermediate Casing Strings

Production casing. Also called the "long" or "oil" string, this string is set through the producing zone and cemented (again tying into the string above). At this point the hole is completely secured with a continuous sheath of pipe and cement isolating it from the formations. The pipe and cement will then be perforated to gain access to the producing zone (FIGURE 8–2, V).

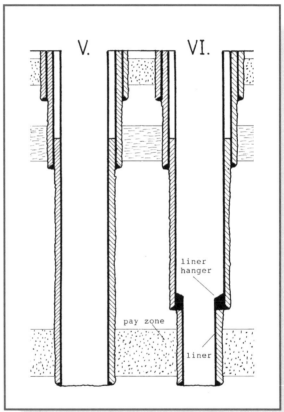

FIG. 8–2   Production Casing and Liner

Liner. When there is concern about the drilling mud causing damage to the producing zone, the long string is often set on top of the zone and the mud replaced with clean fluids to drill out the zone. The zone can then either be completed "open hole" or a liner can be run, cemented in place and perforated. Liners do not extend all the way to the surface like conventional strings. Instead, they are hung from the long string (FIGURE 8–2, VI).

It might seem reasonable to save money by installing the long string as a liner with a much shorter length of pipe hanging from the intermediate string. This is rarely done, however, because of the difficulty of running tools past the restriction of the liner hanger. The convenience of a smooth conduit from top to bottom is universally considered to be worth more than the money saved by the liner configuration.

It's apparent that the pipe size becomes quite small after several strings are run. The casing program must therefore be planned before drilling begins so that the surface hole and pipe are large enough to accommodate all the strings that will be needed, while assuring that the final string is large enough to work in.

Casing program offshore

In offshore drilling, an additional large–diameter string, called conductor casing, is the first string run. When bottom–supported units are used, the conductor is hammered, jetted, or drilled into the seafloor and cemented. It extends to the surface, just below the rig floor, providing a conduit for the drillpipe and mud.

With floating units, a guide structure is affixed to the ocean floor with piles. The blowout preventers then latch into the guide structure on the seafloor and are connected to the surface via a marine riser that is either buoyant or supported by cables. The riser functions as a conductor string.

Casing design

On deep, high–pressure wells, casing design can be critical. If the casing fails, it hazards a multi–million dollar investment; but if it is over–designed, a great deal of money is wasted. Care is therefore taken to optimize the pipe selection and to use the cheapest pipe that is safe.

Once the diameters of the various strings are decided, the drilling engineer determines the optimum wall thicknesses and yield strengths for each string of pipe. This involves extensive calculation, but fortunately computer programs are available to simplify the process. Since it's not feasible for suppliers to carry inventories of all possible pipe specs, the final selection of pipe involves some compromise with what is available from the supplier. The four design criteria for casing strings are:

- *Tension.* As the casing is being run, it all hangs from the top joint. It's therefore often necessary to strengthen the upper part of the string with thicker–walled or higher–grade pipe. If a thicker wall is used, the calculation has to be redone to adjust for the added weight. The buoyancy effects of the mud–filled hole are generally ignored in the calculation, adding an additional safety margin.

- *Collapse.* The danger of collapse is greatest when cement has been circulated up the outside of the pipe because cement is much heavier than the mud that is inside the pipe. The deeper the hole, the greater this outside–in pressure differential, so strengthening for collapse is done at the bottom of the hole.

- *Burst.* As with collapse, burst stresses are concentrated at the bottom of the hole. The critical times are pumping operations at the beginning of a cement job and during a fracture treatment.

  The string is therefore designed to be stronger at the top for tension and at the bottom for collapse and burst, with weaker pipe in the middle of the string.

- *Corrosion.* In deep wells, higher grade pipe must be used to get the necessary strength because thick–walled pipe is too heavy in tension. Unfortunately, higher grade pipe is more susceptible to corrosion. The presence of hydrogen sulfide gas is particularly troublesome because H_2S penetrates high strength steel and embrittles it. In very deep, high–pressure, sour gas wells, nickel–alloy pipe is often the only solution, but it is extraordinarily expensive.

Primary Cementing

Primary cementing is the initial cementing of casing strings in the hole. Cementing done later to repair the primary job or in connection with a workover is called squeeze cementing. The objective of primary cementing is to create a sheath of hard cement completely filling the annular space between the outside of the casing and the hole, thereby blocking fluid movement and pressure transmission up or down the annulus.

Casing equipment

The casing is run with the following accessory equipment (FIGURE 8–3):

- A *guide shoe* with a rounded base is run on the bottom of the string to prevent it from sticking on ledges. The shoe is easily drillable.

- A *float valve* is usually run, either in the shoe (float shoe) or in a float collar located a joint or two above the shoe. This ball–and–seat type valve closes while the casing is being run, preventing mud from filling the pipe. This provides buoyancy to the pipe that lessens the load on the derrick and on the top joint of pipe. As it is run in the hole, the pipe is periodically filled with water at the surface to reduce differential pressure that might collapse the casing. The float valve opens while the cement is pumped, then closes when the pumps shut down, preventing back–flow.

- *Scratchers*, also called wall cleaners, are attached to the outside of the pipe to remove mud cake from the sides of the hole, thereby obtaining better contact between the cement and the formation. If radial–type scratchers (Figure 8–3) are used, the pipe is reciprocated before and during cementing. If vertically–mounted scratchers are used, the pipe is rotated.

- *Centralizers* are attached to the outside of the pipe to center the pipe in the hole during cementing. Centralization is extremely important in assuring the cement sheath completely surrounds the pipe.

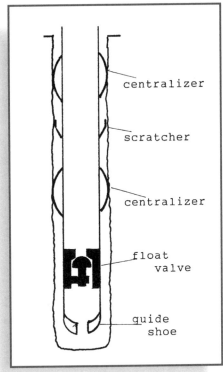

FIG. 8–3   Casing Equipment

Cement mixing

In a continuous process, cement is blended with water, and the resulting slurry pumped down the casing. The slurry is then displaced out of the casing and up the annulus by mud pumped down behind it. The dry cement consists of neat cement plus additives to adjust its properties. This mixture is prepared at a bulk plant before transport to location.

Some of the more common additives are

- *Accelerators* speed up the setting time of the cement

- *Retarders* prevent premature setting in deep, high–temperature wells

- *Density* adjusters increase the cement density to reduce pumping pressures or to permit a higher cement column without fracturing the formation

Cementing operations

Before the pumping of cement starts, the hard rubber bottom plug is inserted into the casing to minimize contamination of the cement by mud. The pump pressure forces the ball off its seat in the float valve, allowing the cement to displace the mud down around the shoe and up the annulus (FIGURE 8–4).

When the bottom plug lands on the float valve, its thin rubber disk ruptures. Pumping continues as the cement is pumped around the shoe and up the annulus. When the last of the cement has been pumped, a solid rubber (no rupture disk) follow plug is inserted into the casing and mud is pumped in behind it.

When the follow plug lands—signaled at the surface by a sharp pressure increase—the cementing job is over. The pumps are shut down and pressure drops. This seats the float valve, preventing the heavier cement in the annulus from U–tubing back into the casing.

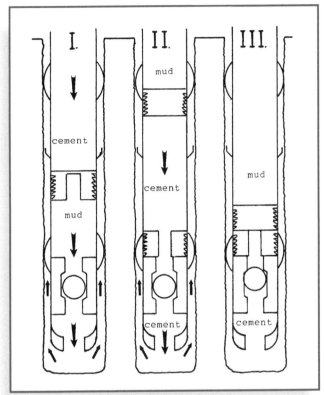

FIG. 8–4 Cementing Operations

Perforating the Casing

In the early days of the industry, the practice was to set casing just above the pay zone, then drill through the pay and make an open–hole completion. The well was then "shot" by detonating nitroglycerin opposite the pay. The explosion removed near–wellbore pay that had been damaged by drilling fluids, exposing a fresh surface for optimal production. The disadvantage of this system was its inability to access a reservoir selectively. For example, it is often useful to complete a reservoir in its relatively thin oil zone while avoiding its water and gas zones.

The current system, with cemented and perforated casing set through the pay, permits selectivity. Bullets were initially used for perforating, but have largely been displaced by the deeper–penetrating shaped charges—also

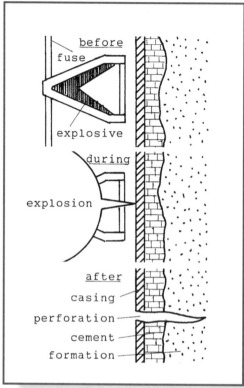

before
fuse
explosive
during
explosion
after
casing
perforation
cement
formation

FIG. 8–5   Jet Perforating

called jet perforators. Perforation densities of four to eight jet holes per foot of pay are typical, with each shot rotated 90° or 180° from the shot above (FIGURE 8–5).

Casing gun

Casing guns are run on electric wirelines and retrieved after firing. They are strongly constructed of steel and can be reused. Easily punctured, soft–metal caps protect the shaped charges while the gun is run in the hole. The ignition wiring is inside the hollow core of the gun. Casing guns leave virtually no debris in the hole.

In the overbalanced condition shown in Figure 8–6, the hole is filled with saltwater—called a water blanket. When the well is perforated, the water rushes out through the new perforations, "killing" the well and preventing a blowout.

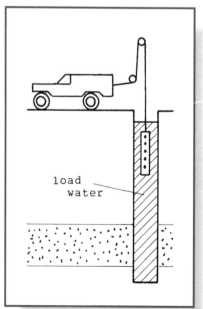

load
water

FIG. 8–6   Perforating with a
Casing Gun

Underbalanced perforating

When the guns fire in overbalanced perforating, the wellbore fluid pushes out into the formation, flushing the jet–charge debris along with it. This can damage the formation in the immediate vicinity of the wellbore and seriously limit the well's productivity. It is therefore better practice to perforate without a water blanket, or "underbalanced". The fluid level in the casing is kept low so that its hydrostatic pressure is less than the formation pressure. The perforating gun is run into the well through an extension of the casing called a lubricator (FIGURE 8–7, I).

When the guns fire in underbalanced perforating, the formation fluids rush into the wellbore flushing out the jet–charge debris and cleaning up the perforations. The stuffing box holds pressure while the well is flowed to clean up (FIGURE 8–7, II). When the gun is pulled back into the lubricator, the lubricator valve on the tree is closed, the pressure is bled off the lubricator, and the lubricator and gun are removed (FIGURE 8–7, III).

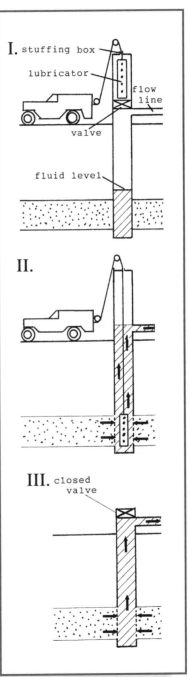

FIG. 8–7   Perforating Underbalanced

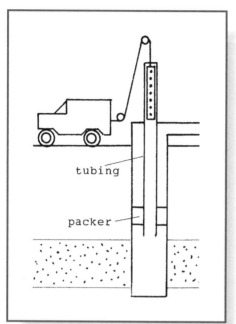

FIG. 8–8   Through-tubing Perforating

Through–tubing perforating

Through–tubing guns are small in diameter, making them easier to run inside of the production tubing. Since the well is already equipped with packers, tubulars, and surface valving, the zone can be perforated underbalanced and immediately placed on production. This avoids any formation damage caused by killing the well to run downhole equipment (FIGURE 8–8).

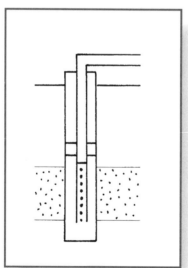

FIG. 8–9   Tubing-conveyed Perforating

Tubing–conveyed perforating

Tubing–conveyed perforating guns are run into the hole below a packer on the bottom of the production string. Like the wireline–run through–tubing system, this system permits underbalanced perforating in a fully equipped well. The advantage over the through–tubing system is that a much longer gun with larger charges can be used. Also, tubing conveyance can push the gun into highly deviated or horizontal holes that would be inaccessible to a wireline–conveyed gun (FIGURE 8–9).

Well Stimulation

When a highly permeable formation is perforated, it may produce freely without a stimulation treatment. This is called a natural completion. Most wells, however, are stimulated either by pumping acid into the formation or by hydraulic fracturing. Stimulation treatments can be highly effective, with production rates doubling or quadrupling. In fact, many profitable producing fields would not be commercial without stimulation.

How does stimulation work?

Stimulation treatment provides the reservoir fluids with better access to the wellbore. This is helpful because in the fluid's journey from the perimeters of the reservoir to the stock tank, there is a major obstacle. This obstacle is the last few inches of reservoir rock that the fluid must pass through before reaching the wellbore.

The geometry of radial flow is illustrated by Figure 8–10 where the flow arrows come closer together as the fluid approaches the wellbore. The fluid interference from this constriction reduces the fluid volume that reaches the wellbore. In high permeability reservoirs, there may be very little restriction, but in tight (low permeability) or moderately tight reservoirs, it is a serious problem.

FIG. 8–10 Radial Flow and Formation Damage

The near–wellbore zone is also critical because formation damage occurs there. The damage is a permeability reduction caused by the formation's contact with the drilling mud. Two types of damage occur:

1. Some formations contain clays that absorb mud filtrate and expand, thereby plugging permeability

2. Solids in the mud become entrapped in the pores of the formation and reduce permeability

Stimulation and profit

The near–wellbore bottleneck is the only restriction on production rate. Upstream of it, the entire reservoir is available for flow. Downstream of it, large enough equipment can be installed in the wellbore to lift any volume to the surface. The well's production rate is therefore seen to be largely dependent on the success of the stimulation treatment.

The production rate—barrels per day (or MCFPD)—is exceedingly significant to the economics of the well. In fact, it is the principal determinant of a well's value. The barrels of proved reserves available for the well to produce is also important, but because of the time value of money, profitability depends more on how fast the reserve is produced than on how big it is. Since barrels per day is the principal measure of a well's quality, it is clear that stimulation treatment is of utmost importance.

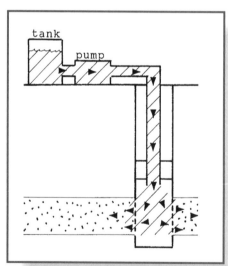

FIG. 8–11 Treating Operations

Treating operations

Stimulation operations consist of pumping treating liquids out of a surface tank, down the well through tubing anchored by a packer, and then out into the formation. Fracture treatments involve several thousand pounds of surface pressure. In onshore operations, the pumps and tanks are truck–mounted. Offshore they are skid–mounted (FIGURE 8–11).

Matrix acidizing. Matrix acidizing is designed to remove formation damage, thereby improving the permeability of the near–wellbore formation. Sandstone reservoirs are normally treated with hydrofluoric acid and limestone reservoirs are treated with hydrochloric acid. The acid is pumped slowly out through the matrix of the reservoir, taking care not to exert enough pressure to fracture the reservoir.

Fracture acidizing. Fracture acidizing is used to stimulate production in limestone and dolomite reservoirs. These rocks are composed largely of calcium carbonate (CaCO3), which dissolves in hydrochloric acid (HCL).

The treatment consists of injecting HCL at high enough pressure to fracture the formation. The orientation of the fracture is roughly vertical, but may deviate from the vertical to follow pre–existing fractures. As the pressure of the pumped acid extends the fractures, it chemically etches an irregular surface on the sides of the fracture. When the pumps are shut down, the fracture closes back up but does not completely heal. The material removed by etching leaves a high–volume flow channel to the wellbore. The fracture changes the flow pattern around the wellbore from radial flow to much higher volume lateral–flow pattern (FIGURE 8–12).

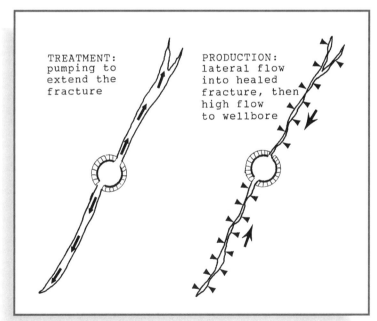

TREATMENT:
pumping to
extend the
fracture

PRODUCTION:
lateral flow
into healed
fracture, then
high flow
to wellbore

FIG. 8–12 Fracture Acidizing

Hydraulic fracturing. Hydraulic fracturing is the most effective stimulation treatment for the tight sandstones typically encountered in older, more consolidated continental sediments. Unconsolidated sandstones typical of coastal sediments normally have very high permeability and do not need stimulation.

Several thousand pounds of surface pressure are usually needed to break down the formation. The fracturing fluid is typically gelled water, using polymers to limit leak–off into the surrounding formation. This keeps sand–face pressures high enough to extend the fracture hundreds of feet. The gel also makes the water slippery, which reduces friction loss, lowering horsepower requirements.

Once the fracture is extended far enough, a propping agent—often large rounded sand grains—is introduced into the gelled fluid being pumped. The gel transports the proppant in suspension as it is pumped out into the fracture. Fluid leak–off into the formation filters the proppant out on the fracture face, packing the fracture. Pumping is then stopped. As pressure dissipates, the fracture starts to heal but is held open by the proppant (FIGURE 8–13).

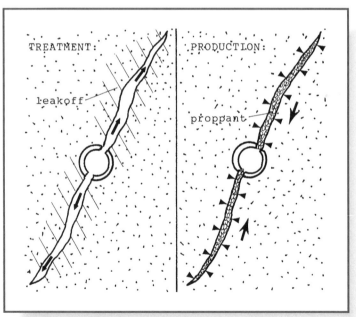

FIG. 8–13   Hydraulic Fracturing

When the well is brought back on production, the fluids move laterally from the formation into the fracture, then flow freely through the matrix of proppant to the wellbore.

Sand Control

Sandstone reservoirs in many producing provinces around the world consist largely of unconsolidated sand grains with virtually no cementation. The permeability is very high, so there is no need for stimulation, but these reservoirs often have a severe problem with formation sand being produced along with the reservoir fluids (FIGURE 8–14).

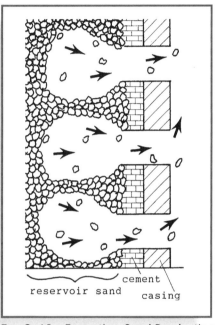

FIG. 8–14   Formation Sand Production

Problems caused by sand production

Sand produced with high–velocity oil and gas streams erodes any steel that it impinges on. This can be a serious problem in a well that produces from two separate reservoirs (dual completion) through two tubing strings (FIGURE 8–15).

Sand entering the wellbore from the upper zone impinges on and cuts through the tubing serving the lower zone. This effectively commingles the two reservoirs, losing reserves by cross–flow from the higher–pressure reservoir to the lower–pressure reservoir. To protect against sand, blast joints of tubing are often run opposite the perforations. Blast joints have extra–thick walls or hard rubber coatings on the outside to help withstand the sand's impact.

Sand erosion is particularly severe where the flow stream

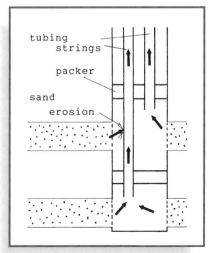

FIG. 8–15   Tubing Erosion by Sand in a Double Completion

changes direction. The well's valve manifold on the surface (Christmas tree) is therefore another danger point. It's not unusual for an elbow fitting on the tree to cut out, which can cause a blowout.

Sand accumulation. Produced sand frequently settles out and bridges over in the well's tubing, completely plugging off production. It's then necessary to work over the well to wash out the sand. Sand also drops out of suspension and accumulates in surface equipment, such as separators, requiring periodic clean–out.

Sand consolidation. Sand consolidation is one approach used for sand control. It usually is done by pumping a resin into the surrounding formation to bond the sand grains together. Another approach is to pump resin–coated sand. The resin hardens and prevents the sand from moving. A major problem with consolidation treatments is that permeability is lowered, sometimes reducing flow below the economic limit. Because of this, consolidation is not widely used.

Gravel–packing

Gravel–packing is by far the most widely used sand control measure because it is effective in a wide range of conditions and has a favorable success rate. In a gravel–pack installation, the produced fluid enters the production tubing through a wire screen. The gravel, which is simply large, graded sand grains, is placed between the screen and the formation sand. The largest gravel is used that will still cause the formation sand to bridge and not flow through it. The holes in the screen are then the largest that will bridge the gravel (Figure 8–16).

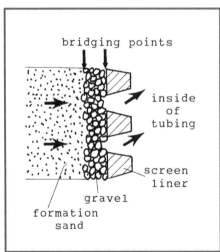

Fig. 8–16   Cross-Section of a Gravel Pack

Gravel packing procedure. Gravel packing is a very expensive procedure, and is successful only 50%–60% of the time. Nevertheless, it is attempted on virtually every completion in sand–producing areas. The procedure is as follows (FIGURE 8–17):

- The wire–wrapped screen is run into the hole on tubing with a specialized "crossover packer" which allows fluids to cross over from the tubing to the annulus, and vice versa.

- The gravel is mixed with gelled water and pumped down the tubing, through the crossover passage in the packer into the annular space outside of the screen. The gelled water passes through to the interior of the screen, crosses over in the packer to the annulus, and circulates to the surface. The gravel is filtered out by the screen, drops to the bottom and accumulates.

- The completed gravel–pack allows sand–free production of reservoir fluids.

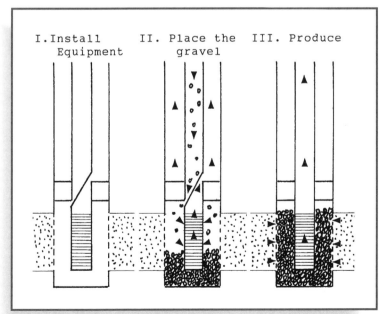

FIG. 8–17 Gravel-packing

Frac–packing

This is a relatively new approach to sand control that blends gravel packing techniques with hydraulic fracturing techniques. The packing fluid is injected at a rate high enough to build up pressure and fracture the formation. Pumping continues as the gravel is packed into the fracture.

Equipping the Well to Produce

The well's casing is rarely used as the conduit for production. Instead, the well produces through smaller diameter pipe, called tubing, that is usually run in the hole with a packer on the bottom. The packer seals off between the tubing and the casing, protecting the casing from the pressure and corrosivity of the produced fluids. In older fields where pressure has declined, tubing is often run without a packer, although a tubing hold–down—a packer minus the sealing element—may be run on beam pumped wells to prevent reciprocation of the tubing (FIGURE 8–18).

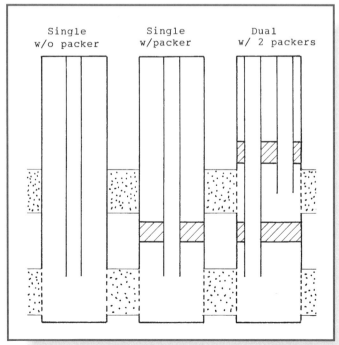

FIG. 8–18   Most Common Completion Configurations

Tubing

Tubing comes in 30–foot joints with threaded couplings. Its diameter is as small as 2–3/8 inches outside diameter (OD) for shallow, low–productivity wells, or as large as 6 inches OD for high–volume gas wells. Tapered tubing strings are often used in deep gas wells with the larger pipe near the surface.

Since tubing strings are relatively light in weight, they can be run in and out of the hole by workover rigs having smaller hoisting equipment than drilling rigs. The strings hang from the tubing hanger in the wellhead and, unlike casing strings, are retrievable.

Packers

Packers come in many configurations, but all have three things in common:

1. A flexible rubber sealing element that closes off the space between the outside of the tubing and the inside of the casing

2. Mechanical dogs that dig into the casing to prevent the packer being pumped up or down the hole by pressure differentials

3. One or more vertical penetrations

There are two main types of packers:

1. *Permanent packers* are run and set on tubing or a wireline, and set with a small explosive charge. The production tubing string(s) is then run through the packer and its rubber sealing elements on the outside seal against the smooth inside bore of the packer. Permanent packers cannot be retrieved but are constructed of materials that can be readily drilled out.

2. *Retrievable or temporary packers* are run integral with (screwed directly into) the production tubing string. They are set variously by rotating, picking up, setting down, or pressuring up the tubing and are designed to be fully retrievable. At times, however, they can be difficult to release.

Tubing accessories

The following are some accessory devices run on tubing strings.

1. *Seating nipples* are constrictions in the tubing used to latch various tools, such as through–tubing bridge plugs, downhole chokes, beam pumps, etc.

2. *Blast joints* are either extra–thick or rubber–coated joints of tubing run opposite casing perforations to prevent sand erosion (see Sand Control)

3. *Sub–surface safety valves (SSSVs)* shut off the tubing automatically in case of an emergency at the surface

4. *Sliding sleeves* are operated by wireline to open or close ports connecting the tubing to the annulus

Dual completions

Dual completions, where two reservoirs are produced simultaneously, but separately, are very common in the oilfield. The economic incentive is that only one well, rather than two, has to be drilled to produce the two reservoirs. By using two strings of tubing and two packers, the two zones can effectively be kept separate and produced with different pressure regimes.

Triple, or even quadruple completions, are sometimes installed. Most companies, however, avoid them because the mechanical complexity of multiple packers and tubing strings, makes it difficult to prevent communication between zones.

Wellheads

The wellhead is mounted on the surface casing. Each additional casing string and the tubing string are then hung from the wellhead as they are run. The wellhead also seals off the annular spaces between the strings. Once the well is completed, operating personnel no longer need access to the wellhead. Onshore wellheads are therefore normally placed just below ground level, either buried or in cellars (FIGURE 8–19).

Christmas trees

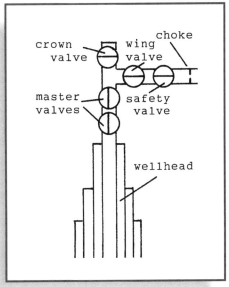

In high–pressure flowing wells, the Christmas tree is the valve manifold that controls flow in the tubing. The tree is strongly constructed to contain full reservoir pressure. The functions of the various valves are as follows:

FIG. 8–19   Wellhead and Christmas Tree

- The *master valves* are used to shut in the well. Many trees have dual master valves with only the top valve used while the bottom valve stays open. Should the top valve sustain wear and start leaking, the bottom valve is always in good condition to control the well while the top valve is repaired.

- The *crown (or lubricator) valve* is used when a lubricator is attached to perform through–tubing well service.

- The *wing valve* is normally used for routine opening and closing of the well.

- The *choke* is an orifice that is varied in size to control the well's flow rate. It also protects downstream equipment by confining full well pressure to the tree.

- The *safety valve* automatically shuts in the well if unsafe conditions occur, such as excessively high or low downstream pressure.

FIELD APPRAISAL AND DEVELOPMENT 9

Onshore Development

When an onshore wildcat well encounters what appears to be commercial hydrocarbons, the well is quickly completed and placed on production.

Production testing

Unless there has been a flowing drillstem test on the well, there is considerable uncertainty as to what the well's production rate will be. After a few days of production to recover the treating fluids and clean up the well, the Initial Potential Test (IP) is taken. The IP is of great interest, so it is carried on the morning drill report.

The IP data includes the barrels of oil and water and the cubic feet of gas produced in 24 hours. If the well is flowing, the tubing pressure upstream of the choke is recorded along with the choke size (in 64ths of an inch). This is important in judging the potential of the well. If the well is under pump, the dimensions of the pumping system are reported. Once the IP has been reported, the well is usually taken off the drill report.

Good IPs can, however, be misleading. Completions in some types of reservoirs come in at a high rate initially, but then decline precipitously over the next few weeks—sometimes to a degree that renders further development of the reservoir uneconomic. It's therefore prudent to produce a new discovery well for some time before embarking on an ambitious drilling program.

Gas reservoirs are particularly unpredictable, and some very costly mistakes have been made. When a gas discovery is made some distance from a pipeline, a number of wells may need to be drilled before adequate gas reserves are established to justify laying a pipeline to the field. Since testing the production rate of these wells requires venting the produced gas to the atmosphere, the test period is held to a minimum —often less than 24 hours. In some cases this is not long enough to accurately assess the extent of the reservoir. Oil wells don't have this problem because, unlike gas, test oil can be trucked from the location.

Step–out drilling

Once the discovery well is completed and put on production, development drilling begins. The act of picking drilling locations is very different than it was during the exploratory phase and requires differently trained people. The exploration geologist was an expert in picking wildcat well locations in areas where virtually no subsurface data was available. In development drilling, however, a great deal of data is available from the cores, sample logs, drill-stem tests, and electric logs obtained from the discovery well. The development geologist that now picks the locations is expert in working with this data. The process is step–out drilling (FIGURE 9–1).

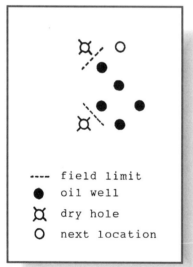

---- field limit
● oil well
⌀ dry hole
○ next location

FIG. 9–1   Step–out Drilling

In developing structural traps, usually defined by seismic, the process is to step out one location at a time, drilling as high on structure as possible (FIGURE 9–2). This assures that each subsequent well is drilled in the spot with the highest probability of success. As development proceeds downdip, an oil–water contact may be encountered which defines the outer limits of the field. The remainder of the reservoir above this contact is then developed by stepping out— again one well at a time. Caution is appropriate because reservoir quality— porosity, permeability, or thickness—could disappear at any point. Dry holes are a normal part of every field's development, but every effort is made to minimize them.

When developing stratigraphic traps, seismic is of little help, so each step–out is a potential dry hole. Every bit of available information is used to improve the odds. For example, if existing fields in the area show a southeast–northwest orientation, the geologist would probably have the initial stepouts follow that pattern.

Unlike wildcat wells, the majority of development wells are successful. However, when probing a field's limits, some dry holes are inevitable—particularly when dealing with stratigraphic traps.

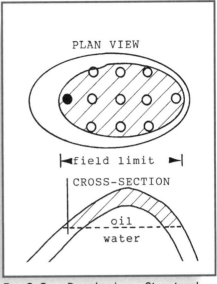

FIG. 9–2   Developing a Structural Field

Well spacing

A major issue in drilling development wells is how far apart they should be. That is, how many wells are needed to deplete a given reservoir?

Given a continuous reservoir and adequate time, a single well would eventually produce all the reserves—but it could take two or three hundred years! The time–value of money therefore comes into play, dictating that additional wells be drilled. The economic rationale is that additional wells accelerate revenue, and should be drilled if the return from investing that accelerated revenue exceeds the cost to drill them.

By making assumptions about future costs, prices, and interest rates, the theoretical optimum spacing can be calculated for each field. In the United States, for instance, oil well spacings are typically 40, 80, or 160 acres (An acre is 43,560 square feet). Since fluid viscosity is a major determinant of spacing, heavy–oil fields may be drilled as tightly as one–half acre per well, whereas gas fields typically use 640 acre spacing (FIGURE 9–3).

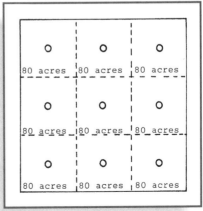

FIG. 9–3   80–Acre Spacing

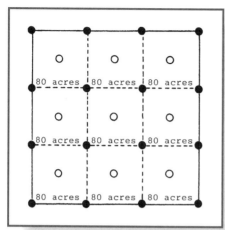

FIG. 9–4   Infill Drilling Locations

Infill drilling

After a field has been developed and produced for some time, the decision is often made to drill infill wells (FIGURE 9–4). If, for example, the field was originally drilled on 80 acre spacing, an infill well is drilled in the center of each square area formed by four wells. This doubles the number of wells and reduces the spacing to 40 acres.

Reasons for infill drilling are:

- The reservoir may have proven to be so tight (poor permeability) that it cannot be depleted within a reasonable time using the original spacing. Infill wells encounter reservoir pressure close to original values and produce at relatively high rates.

- The reservoir may have proven to be non–continuous, made up instead of numerous unconnected lenses. Some of the lenses may not have been penetrated on the original spacing and remain undepleted. Infill wells would tap these lenses and produce enough to justify the drilling costs (FIGURE 9–5).

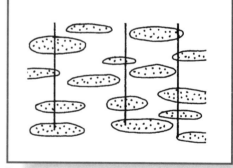

FIG. 9–5   Infill Wells Needed to Tap all Lenses

- Secondary and tertiary recovery projects require that as many as 50% of the field's wells be used for injection. Injectors can be either converted producing wells or new drills. In either event, infill drilling usually takes place.

Over–the–Water Development

The water depth and the intensity of wind and waves determine the equipment and procedures appropriate for each location.

Shallow, protected waters (marshes and estuaries)

Development proceeds similar to onshore development with the drilling rig moving from location to location drilling and completing vertical holes. The rig is mounted on a barge which is pushed to the location and flooded until it sits on bottom (FIGURE 9–6). After the well is drilled and completed, the barge is re–floated and moved to the next location, leaving the wellhead and christmas tree protruding out of the water.

FIG. 9–6   Barge–mounted Rig

A work barge is moved in and a small jacket, or "toadstool" set down over the tree and pinned to the bottom with piles driven through its legs. It functions as a dock for the small boats used by the operating personnel, provides a place for operating personnel to stand while working on the tree and protects the wellhead from being accidentally struck by boats, barges, or floating debris (FIGURE 9–7).

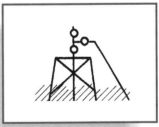

FIG. 9–7   Tree and Jacket

When the water depth is less than 10 or 12 feet, dredges are used to cut access canals for the drilling barges. This system is effective into very shallow water, and even onshore in low–lying swampy areas.

Moderate water depths

When water depths reach 25 to 30 feet, jackup drilling rigs are used for both exploratory and development wells. The wells are straight holes (not deviated) with their surface locations in a grid similar to that used onshore

FIG. 9–8 Jackup Drilling Rig with Derrick Cantilevered over Platform

and in protected waters. Each well has a piled jacket that is either set down over the tree after completion or pre–set and drilled through from the jackup's cantilevered floor (FIGURE 9–8).

Deeper water, moderate sea states

As the water gets deeper, platform costs escalate sharply and shift the economics away from vertical wells with individual jackets toward multiple directional wells drilled from a single platform. Larger fields will have multiple platforms.

The discovery wildcat may be drilled with either a semi–submersible or a jackup—some jackups can operate in over 400 feet of water. Wildcats are usually drilled as expendable holes—to be filled with cement and abandoned after logging—since it's unlikely the future platform would not be set over their location. Additional expendable holes may then be drilled as appraisal wells to delineate the reservoir. This allows optimal placement of the platform and planning of the development drilling program.

A multi–well platform designed for the precise water depth and bottom conditions is then constructed, installed, and a drilling rig set on it. All the facilities associated with the rig may be located on the platform, or the rig might be tender–supported (FIGURE 9–9). The floating, anchored tender—either ship–shape or semi–submersible—contains crew quarters, mud pits and pumps, cement mixing and pumping equipment, pipe & equipment storage, etc. Umbilical chords carrying power, water, and mud connect the tender to the platform.

FIG. 9–9 Tender-supported Platform Rig

After each well is drilled and completed, the rig is skidded a few feet to the next slot and a new well is spudded. When all the wells have been drilled, the rig is lifted off by cranes on the tender, and the tender is towed to a new platform.

The following safety–connected issues arise in this operation:

- Should the initial wells that are drilled be produced while the later wells are still being drilled, or, for safety reasons, should production be delayed until drilling operations on that plat-form are complete?

- Should the treating facilities (the major source of fire and explosion) be on separate platforms from the wells (the major investment) and the living quarters (the major personnel risk)? The deeper the water, the stronger the economic man-date for a single platform to house all three functions.

Deep water, severe sea states

The appraisal process is much more intensive in deep water. It is necessary to make certain that the field size and the production rate will be high enough to justify the enormous front–end expenditure for a platform. This is a very different economic profile from onshore or shallow–water development where each well stands alone economically. A field discovered in deep water must be quite large to be commercial, whereas even a one–well field can be commercial onshore.

In deep waters, floating semi–submersible drilling rigs are normally used to drill the expendable wildcat and appraisal wells. Drill–ships are also used at times.

A single platform is normally used—either bottom–supported or a floater such as the tension–leg configuration. Having separate platforms for the drilling rigs and wells, the quarters, and the fired facilities is prohibitively expensive in deep water. For example, a bottom–supported platform in the northern North Sea costs more than a billion U.S. dollars.

The high cost and technologic challenges of setting platforms—either floating or bottom–supported—in extremely deep water has increased the use of subsea completions, where the wellheads are installed at the mudline, then connected by pipeline with nearby platforms. This makes particular sense for small fields developed in the vicinity of older fields. The production decline in the older field creates surplus facility capacity that can be used by the new field.

Site Preparation

Onshore

Preparation of an onshore location for surface facilities is relatively simple. The ground is leveled, a drainage system is installed and crushed rock or other surface–stabilizing material is spread. In swamps, it may be necessary to build raised "corduroy" locations and roads paved with planks, but it is still a relatively inexpensive operation.

Protected waters

Preparing locations in shallow, protected waters such as lakes or marine estuaries is considerably more difficult and expensive. Three types of construction are used:

- *Dredged bottom material* may be used to raise the location above water.

- *Wooden or steel piles* may be driven into the bottom to support a wood, steel or concrete deck. Using a pile–driver (which repeatedly picks up and drops a heavy weight on the pile), piles are driven to their "point of refusal" where penetration stops. The load on the pile is supported by the frictional force between the pile and the surrounding earth (FIGURE 9–10).

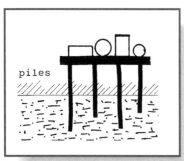

FIG. 9–10   Pile-supported Deck

- *Surface facilities* are often pre–fabricated in the yard and installed on floating concrete

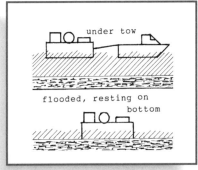

FIG. 9-11 Concrete Barge
Installation

barges. The barges are then towed to the location and flooded to rest securely on the bottom (FIG. 9–11).

Offshore

Bottom–supported steel platforms. In deeper, unprotected waters, bottom–supported steel platforms are constructed to provide the necessary load–bearing space. Steel platforms are built and installed in two sections: the jacket extending from the mud line to just above the water line and the deck section. The installation procedure is as follows:

1. The jacket is loaded onto a barge and towed to location (FIGURE 9–12, I).

2. The barge is partially flooded and the jacket is launched. The structure is caused to float horizontally at the surface by flooding selected jacket members (FIGURE 9–12, II).

3. The jacket is brought slowly to a vertical position by flooding additional chambers, then set down on the sea bottom (FIGURE 9–12, III).

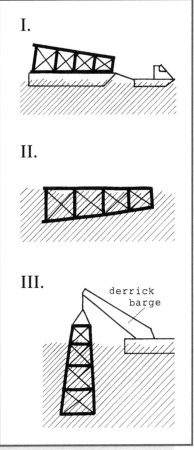

FIG. 9-12 Jacket Tow, Launch,
Set

IV.

piles

FIG. 9–12 CONT'D   Setting the Deck on the Jacket

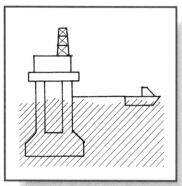

FIG. 9–13   Concrete Platform Under Tow

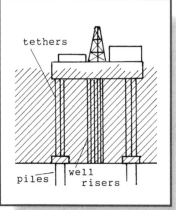

tethers

piles   well risers

FIG. 9–14   Tension Leg Platform

4. Steel pipe piling is driven down through the jacket legs deep into the seabed. The derrick barge then lifts the deck section from a barge and sets it on the jacket (FIGURE 9–12, IV).

5. The final step is for the derrick barge to stack pre–fabricated facility modules on the deck.

Concrete gravity platforms. Concrete platforms have been used extensively in the Norwegian North Sea. They are fabricated at remote construction sites, floated to location, and flooded to sink them to bottom (FIGURE 9–13). Although held in place largely by their weight—hence the "gravity" designation—some also have a few "skirt" piles distributed around their base.

Note that installing a bottom–supported platform, either steel or concrete, creates the obligation to decommission and dispose of the platform when the field is abandoned. This can be very costly.

Buoyant platforms. In very deep water where the cost of bottom–supported jackets becomes prohibitive, buoyant platforms are increasingly being installed. The most popular configuration is the tension–leg platform which is tethered to piled pads on the ocean floor by strings of steel pipe. The platform's movement in severe sea states is controlled by keeping the strings under considerable tension (FIGURE 9–14).

Spars are buoyant vertical cylinders affixed to the bottom and stabilized by anchor lines. Because the configuration minimizes vertical motion caused by surface conditions, spars are increasingly being selected to support production and drilling facilities in very deep water (FIGURE 9–15).

Sea–going vessels, temporarily anchored in place, are increasingly being used in place of permanent platforms. These FPSOS (Floating Production, Storage and Offloading vessels) may be ship–shaped or semi–submersible and are used in conjunction with subsurface completions. The advantages of this system are that it is cheaper than platforms, production can be started quicker, andwhen the fields deplete, FPSOS are easily disconnected and moved to other fields.

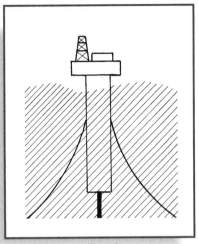

FIG. 9–15   Drilling Spar

Well Workovers

After a well has been producing for a while, it is often necessary to re–enter and repair it. This is variously called a workover, recompletion, reconditioning, or remedial job.

Workover Units

For onshore operations, the units are normally truck–mounted, whereas offshore platform–based units are skid–mounted. For work in shallow water, the units are often mounted on mobile jackup barges. The various types of workover units are listed below.

- *Slick–line wireline units.* Single–strand wire is used for routine maintenance of gas–lift valves, chokes, sliding sleeves, and for cutting paraffin accumulated in the tubing. Slick–line is run through–tubing, under pressure, using a lubricator .

- *Electrical–conducting wireline units.* Multi–strand, electrical–conducting wire is used for logging, perforating, and setting plugs and packers. It is run both through–tubing under pressure and through–casing with the well killed.

- *Coiled tubing units.* Small diameter (3/4" to 1–1/2"), continuous (no joints) tubing is rapidly spooled off a reel into the hole, either through–tubing or through–casing. It is used for many purposes such as cleaning out sand and placing cement and is particularly useful for working in horizontal holes.

- *Pulling units.* These are mobile onshore units using 2" or 2–1/2" OD jointed tubing as a work string. The drawworks is smaller than that used on drilling rigs, usually pulling doubles. A rotary table and reverse–circulating mud system can be attached for drilling out plugs, junk, or for deepening. Pulling units work with the tree removed, so flowing wells have to be killed. They are multi–purpose machines, used for routine pulling of rods and tubing as well as for major workovers.

- *Conventional drilling rigs.* The drilling rigs are often left on deep–water platforms to function as workover units.

Types of workovers

Plugging back to a new zone. Wells may encounter more than one productive zone. Often the bottom zone is completed initially and produced to depletion. The well is then worked over to plug off the old zone and complete a new, upper zone (FIGURE 9–16). If onshore, the workover procedure is to

1. Pump saltwater down the tubing to kill the well.

2. Move a pulling unit on location, remove the tree, and pull the tubing and packers.

3. Run a cement retainer (a type of packer) on an electric wireline, setting the retainer above the old zone.

4. Run a tubing workstring and latch into the retainer. Pump cement down the tubing and then slowly into the pay zone until it pressures up. This is called squeeze cementing.

5. Pick up the tubing and unlatch from the retainer. A flapper in the retainer swings closed to prevent cement back–flow.

6. Pump water down the annulus and up the tubing (reverse circulation) to clean out the cement left in the tubing.

7. Come out of the hole with the tubing. Go into the hole with a perforating gun on wireline and perforate the upper zone.

8. Fracture, acidize or gravel–pack the upper zone as necessary, then rerun the production tubing and packer, install the tree, and bring the well back on production. To unload the kill fluid, it might be necessary to pump liquid nitrogen down a string of coiled tubing run inside the production tubing.

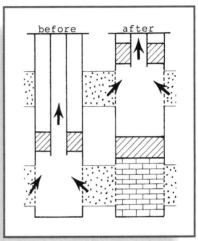

FIG. 9–16 Plugback to a New Zone

Re–stimulation. The original pay may need to be re–acidized or re–fractured.

Deepening to a new zone. The existing perforations are squeezed off; the hole is drilled deeper to the new zone; a casing liner is run and cemented; and the new zone is completed.

Repairing casing leaks. This may involve setting a bridge plug below the leak and a packer on tubing above the leak, then squeeze–cementing the leak. Another approach is to set a "scab liner" over the leak. This is a short piece of smaller–diameter casing either with packers on both ends or that expands outward to seal the leak.

Replacing faulty well equipment. Most workovers simply replace down-hole equipment such as leaking tubing, broken sucker rods, malfunctioning gas–lift valves, or leaking packers.

ARTIFICIAL LIFT   10

Natural Lift

Wells completed in newly discovered oil reservoirs usually flow to the surface by natural lift. Reservoir pressure provides the energy to move reservoir fluids horizontally into the wellbore, then up the tubing and through the surface facilities.

Flowing wells

Reservoir pressure is usually equal to the hydrostatic head of a saltwater column from the reservoir to the surface. This is roughly one–half psi per foot of reservoir depth (see Reservoir Pressure, chapter 2). Figure 10–1 illustrates how this causes flowing wells.

Drawn on the left side of Figure 10–1 is a glass tube bent in a "U" shape and filled with liquid. The base of the "U" contains saltwater, as

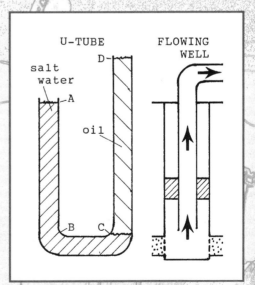

FIG. 10–1 U-tube and Flowing Well

does the left–hand vertical tube AB. The right–hand vertical tube CD contains oil. Since the pressures at points B and C are the same, the fluid column CD weighs the same as AB. However, the pressure supports a taller column of oil because oil is less dense than saltwater.

The U–tube helps explain the flowing well on the right side of Figure 10–1. As with the U–tube, the reservoir pressure is determined by the hydrostatic column of saltwater extending from the reservoir to the surface. That same pressure (C on the U–tube) supports a column of oil in the tubing that extends not only to the surface, but beyond under the following conditions:

- If the tubing is open to the air, the oil "blows out" in a geyser

- If the tubing valve is closed at the surface, there is significant pressure on the tree

- If, as shown, the tubing is piped through production facilities, the well flows oil

Pressure needed to flow. A flowing well must overcome the following obstacles (FIGURE 10–2):

1. *Reservoir friction.* Friction from the horizontal flow of reservoir fluids through the rock matrix to the wellbore causes a pressure drop, or drawdown. The higher the production rate and the greater the viscosity of the fluids, the greater the drawdown.

2. *Hydrostatic head.* The weight of the fluid column in the tubing must be overcome.

3. *Tubing friction.* At high flow rates, there is considerable friction between the produced fluid and the inside of the tubing. This is particularly significant in deep gas wells.

4. *Facilities pressure.* The well must overcome the pressure of the surface facilities. In oil fields, the first vessel is usually a gas separator that carries one or two hundred psi of pressure. The separator pressure then moves the fluids through the treating system. Pressures on gas reservoir facilities are higher.

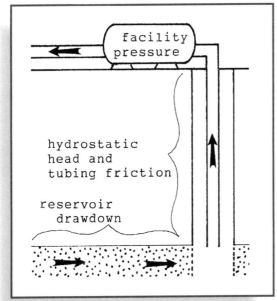

FIG. 10-2   Obstacles to a Flowing Well

Why a flowing well dies

If a well is produced wide open (no choke), production stabilizes at the rate where reservoir pressure is exactly offset by the sum of the four obstacles previously mentioned.

As the reservoir pressure depletes with production, the friction across the formation and through the tubing drops the pressure further, so the production rate drops.

When reservoir pressure drops to the sum of the hydrostatic head and the facility pressure, it can bear zero friction loss, so production stops. The well has "died" and artificial lift measures are required to continue production.

Solution gas effects

Bubbles of gas break out of solution and mix with the oil flowing up the tubing. As the bubbles rise toward the surface, the hydrostatic head reduces, which multiplies the bubble's size. The bubbles assists oil flow in two ways:

1. The expanding gas in the tubing displaces the much denser oil, which lightens the overall fluid column; this increases the velocity and rate of flow

2. The buoyant gas bubbles push up through the oil in the tubing with a piston–like effect that accelerates the oil's upward movement

The more gas the oil has in solution, the more lifting help is provided. Therefore, reservoirs with volatile crudes containing more light ends flow longer, depleting reservoir pressure further than reservoirs with less volatile crudes.

Produced water effect

Since saltwater is heavier than oil, its presence in the production stream increases the hydrostatic head of the fluid column, reducing the production rate.

If the producing stream is 100% saltwater, both legs of the U–tube in Figure 10–1 are filled with salt water. The fluid level rises only to the surface and the well is dead.

Depending on the water saturation and relative permeability of the reservoir, wells in newly discovered reservoirs (virgin pressure) may produce water–cut oil, and may or may not flow depending on how high the water–cut is. Wells that flowed initially may later die as the water–cut is increased by encroaching water.

Gas wells

Since the density of gas is a small fraction of the density of even the lightest oil, the hydrostatic head of gas in tubing is minimal. As a result, dry gas fields continue to flow until reservoir pressure approaches separator pressure.

In summary

Whether an oilwell flows, and how long it flows before artificial lift is required, depends on reservoir pressure, oil volatility, and producing water–cut.

Sucker–Rod Pumping

System components

Prime mover. The prime mover provides the power to operate the pumping system. If electric power is available, electric motors are the most convenient prime movers for rod pumping. Field–gas fired engines are the alternative. The prime mover's power is transmitted to the pumping unit via flexible V–belts.

Beam pumping unit. The pumping unit transforms the rotational motion of the prime mover to the vertically reciprocating motion of the sucker–rods. The beam–type pumping unit shown is the most widely used. Its operational sequence is

1. The **V–belt** from the prime mover rotates the gears

2. The gears in the **gear box** moves the back end of the walking beam up and down

3. The **walking beam** pivots on the samson post, moving the horsehead up an down

4. The **horsehead** reciprocates the sucker–rod string vertically

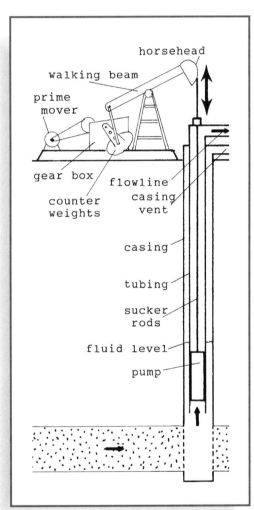

FIG. 10–3   Sucker Rod Pumping System

The counterweights save power cost by distributing power usage evenly over the up and down strokes. On the downstroke, the weight of the dropping rods lifts the weights into position; then on the upstroke the weights drop, lifting the rods and fluid column.

Sucker–rod string. The sucker–rod string is run inside the tubing and attaches to the pump plunger. The rod's reciprocation strokes the plunger up and down in the pump barrel, lifting the produced fluids up the tubing.

Sucker rods are manufactured in either 25 or 30 foot lengths, have diameters from 1/2 inch to 1–1/8 inch and have screwed couplings. Most rods are made of steel, but fiberglass–reinforced vinyl is used in some corrosive applications.

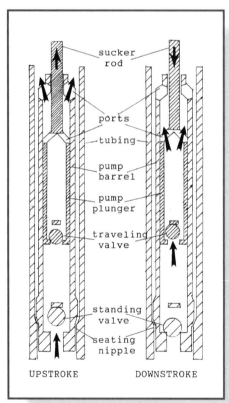

FIG. 10–4 Sucker Rod Pump

Downhole pump. Sucker–rod pumps come in many configurations. Most are run into the hole on the rodstring and latch into a seating nipple in the tubing.

The pump in Figure 10–4 consists of two telescoping cylinders—the plunger and barrel—and two ball–and–seat valves—the traveling and standing valves. The plunger fits closely inside the barrel with a metal–to–metal seal.

When the rod string lifts the plunger, the ball in the traveling valve seats and seals. As the plunger rises, it forces the fluid in the upper part of the barrel out through the ports into the tubing. This lifts the entire fluid column in the tubing upward, causing production to spill over at the surface into the flowline. The rising plunger also causes the formation fluids to lift the ball in the standing valve and refill the barrel.

As the plunger starts downward, the traveling valve opens and the plunger drops freely through the fluid in the barrel. The standing valve closes, preventing fluid from being pushed out through the bottom of the barrel.

Pumping rate

Since the rate of fluid entering the wellbore constantly changes as reservoir pressure drops, waterflooding increases reservoir pressure, and other factors have effect, it's necessary to periodically change the pumping rate of the system.

Annular fluid levels. If the pumping system is lifting all the oil that enters the wellbore, the fluid level in the well's casing/tubing annulus stays at or near the pump. The operator is interested in maintaining this pumped–off condition since it recovers the maximum fluid.

If the pump is not keeping up with the fluid influx, the annular fluid level rises. Operators therefore regularly check fluid levels using acoustic devices, and increase the pumping rate if high fluid levels are found.

Fluid pound. It is also undesirable for the pumping rate to exceed the fluid influx. If the pump barrel doesn't completely fill with liquid on the upstroke, the plunger returning on the downstroke slaps the liquid surface sharply. This fluid pound sends out shock waves that damage balls, seats, and rods.

Adjusting pumping rate. The output of the pumping system can be adjusted to eliminate high fluid levels or fluid–pound as follows:

- The pumping unit's strokes–per–minute can be increased or decreased by changing the size of the V–belt sheave (pulley) on the gear box

- The length of the pump stroke can be changed by using a different hole in the pumping unit crank

- A smaller or larger diameter pump can be run

- If the prime mover is an electric motor, a time–clock can automatically turn the pumping unit on and off as needed to optimize pumping time

Pump–off controllers. The immediate problem can be solved by the methods above, but further adjustments will be necessary as reservoir pressure continues to decline, or as enhanced recovery processes start to stimulate a reservoir. Pump–off controllers were developed to automatically adjust to these changes.

These controllers sense the first tremor of fluid–pound and automatically shut off electric power to the pumping unit before any damage is done. Pumping then automatically restarts in approximately 15 minutes, and continues until pump–off is again signaled. The pump–off controllers thereby prevent both high fluid levels and damaging fluid–pound.

Gas production.

The efficiency of sucker–rod pumps is reduced greatly by free gas in the production stream. As the plunger drops, free gas in the barrel tends to compress in place rather than displacing through the traveling valve and on to the surface.

This gas–locking can often be prevented by setting the pump below the casing perforations. This separates the oil from the gas and diverts the gas up the casing/tubing annulus. Another approach is to install a gas anchor on the tubing to separate the gas and divert it up the annulus.

Applications

Sucker–rod pumping systems are extremely dependable and require minimal maintenance. They are therefore the dominant pumping system used onshore, but are not used offshore because of their weight and bulk. They also have depth and volume limitations that dictate alternative pumping systems in some onshore situations.

Gas–Lift

In gas–lift, high–pressure gas is injected down the casing annulus into the tubing string, lifting the fluid by lightening the fluid column, and, by a piston effect accelerating the oil's upward movement. A supply of lift gas is required throughout the life of the project, so most applications involve relatively high

GOR, volatile oil production. Aside from a compressor to bring the gas up to injection pressure, minimal surface equipment is necessary. Gas–lift is therefore practical for offshore operations and, in fact, dominates that application.

System operation

In Figure 10–5, the well on the left side died because reservoir pressure no longer supported a column of liquid all the way to the surface. The well on the right has been equipped for gas lift by installing a packer on the tubing and gas lift valves.

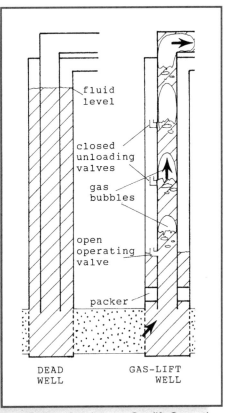

1. High–pressure gas is injected into the annulus at the surface, which displaces the liquid level downward, forcing the liquid into the tubing through the top unloading valve

2. When the gas reaches the top valve, it enters the tubing and unloads the well

3. The liquid level continues downward until gas enters the second unloading valve; the top unloading valve then closes and the well is unloaded only from the second valve

FIG. 10–5 Continuous Gas-lift Operation

4. When the liquid level reaches the operating valve and the well is unloading from that depth, the unloading valves are all closed (there are usually more than the two shown); this is the steady–state operating condition of a continuous gas–lift system

5. The injected gas tends to form large bullet–shaped bubbles trailed by smaller bubbles; as the bubbles rise rapidly up the tubing, the reducing hydrostatic pressure causes them to continuously expand

Valve installation

The valves are mounted in mandrels that are part of the tubing string. There are two types of mandrels (FIGURE 10–6):

- *Conventional mandrels* are run on the tubing with the valves in place; valves can be retrieved only by pulling the tubing

- *Side–Pocket mandrels* are used for retrievable valves; these mandrels are run empty and the valves are then installed by through–tubing wireline; the pocket of the mandrel is sufficiently offset to cause no restriction to through–tubing wireline tools

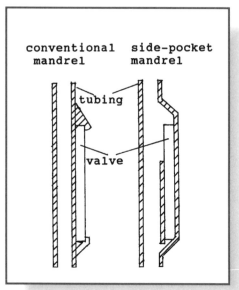

FIG. 10–6   Gas-lift Mandrels

Other configurations

The gas–lift discussed to this point is continuous, tubing–flow lift. Most installations are of this type, but the following other configurations are used at times.

Intermittent lift is sometimes used for low–capacity wells. An intermitter valve on the surface gas line opens for a preset period, injecting a single large slug of gas down the casing, through the operating valve, and into the tubing.

Dual gas–lift can simultaneously lift both tubing strings in a dual complet-ed well. This system is difficult to control.

Annulus flow is used for large capacity wells. The gas is injected down the tubing and through the valves into the annulus.

Applications

Gas–lift works well in high gas/liquid ratio and sand–producing wells. It is unaffected by wellbore deviation and performs well in deep wells. The equip-ment is inexpensive and minimal surface facilities are required.

The disadvantages of gas–lift are that is very high maintenance—requir-ing continual adjustment—and is very inefficient in terms of the work done per unit of energy consumed. Gas–lift is therefore used only in situations where other systems are not practical, such as offshore.

Electric Submersible Pumping (ESP)

ESPs are multi–stage centrifugal pumps coupled to electric motors (Figure 10–7). They are run in the hole on the bottom of the tubing with the power cable strapped to the outside the string. They are widely used in the oilfield for high–volume, shallow production.

System components

- The *electric motor* has an elongated design to fit down the hole; it is a three–phase, squirrel–cage, induction–type, and operates at 2800–3500 revolutions per minute.

- The *sealing section* allows the fluid in the motor to pres-sure–equalize with the downhole fluids while not mixing with them.

- The *pump intake* includes a gas separator that diverts the produced gas up the annulus rather than through the pump; the pump has many stages, consisting of a rotating impeller

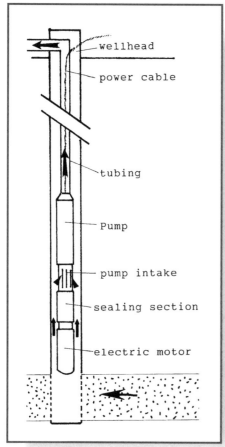

FIG. 10–7   Electric Submersible Pump

and a stationary diffuser, stacked on top of one another and operating in series.

- The **power cable** is heavily insulated and externally armored. Extreme care is taken while running the tubing not to pinch and damage the cable.
- The **wellhead** is small and lightweight, so ESPs can be used offshore.

Applications

ESPs are sophisticated, high–performance devices that are expensive to purchase, repair, and operate. Their use is therefore limited to high–volume applications beyond the capabilities of other systems. Water–drive and waterflood production with high water cuts are typical applications.

Other Systems

Sucker–rod pumping, gas–lift and ESP, in that order, dominate artificial lift. The following other systems are, however, occasionally used.

Power–oil systems

These systems transmit lift energy to the formation face via pressurized power oil pumped down a separate tubing string.

Hydraulic pumping uses power oil to stroke a downhole piston pump. The formation fluids and the returning power oil are pumped up the production tubing.

Jet pumping forces the power oil through an orifice, providing lift to the produced fluids.

Plunger–lift uses power oil to lift a plunger and its load of produced fluids up the tubing to the surface.

Progressing cavity pumps

Downhole progressing–cavity pumps consist of a rotating worm–shaped rotor inside of a flexible stator. The rotor is driven by a rod string inside the tubing, which is rotated by a motor–driven surface unit.

This is a relatively new application for the progressing–cavity system, which is also used for both surface pumps and downhole drilling mud motors. The system is particularly good for pumping abrasive fluids.

SURFACE FACILITIES 11

This chapter covers the surface facilities that handle produced fluids beyond the wellhead. Included are

- *Separation facilities* to separate the oil, water, gas, and gas liquids

- *Measurement facilities* to determine the volumes of oil, water, and gas produced and sold

- *Storage facilities* to store produced liquids pending disposal

- *Transportation facilities* to gather and dispose of produced fluids

Subsurface equipment is closely involved with reservoir performance, so it is normally designed by petroleum engineers. Surface equipment, however, is sufficiently remote from the reservoir. Its design is normally done by mechanical, civil, or process engineers.

Unlike a manufacturing plant which produces at a fixed rate throughout its life, the production rates of reservoir fluid streams vary widely as the field depletes. This complicates the sizing of surface equipment.

Equipment Elements

The complex maze of piping, vessels, tanks, etc. that makes up the typical oil-field surface system can be baffling. These systems become easier to understand when broken down into their elements.

Pressure vessels

The large cylindrical pressure vessels seen in every producing facility are essentially enlarged sections of a pipeline. The enlargement allows fluid velocity to slow, providing the retention time necessary for the fluids to gravity–segregate. Gravity segregation is the primary mechanism used to separate produced oil, water, and gas.

Pumps

Pumps pressurize liquids, creating a pressure differential that moves liquids through the system. Two types of pump are commonly used in the oilfield.

Centrifugal pumps consist of impellers rotating at high speed within stationary cases. The liquid enters the pump at the center of the impeller and is picked up by the impeller vanes and whirled in a circular motion. Centrifugal force flings the liquid outward, creating the pump discharge pressure (FIGURE 11–1).

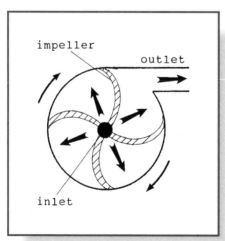

FIG. 11–1 Centrifugal Pump

Centrifugal pumps are best suited for high volume, low pressure applications. They are relatively trouble–free to operate and if a downstream valve is inadvertently closed, they pump through themselves without damage.

Inexpensive single–stage (one impeller) centrifugal pumps, powered by electric motors, are routinely used for moving liquids around petroleum facilities. Expensive multi–stage centrifugal pumps, often diesel engine or turbine powered, are used for larger tasks such as pipeline pumping.

Positive–displacement pumps use pistons reciprocating within closely fitting cylinders to displace the liquid. Pumps consist of a gearbox converting the rotating V–belt drive to reciprocating motion and a fluid end where the pumping takes place. Prime movers are usually natural gas or diesel fueled internal combustion engines (FIGURE 11–2).

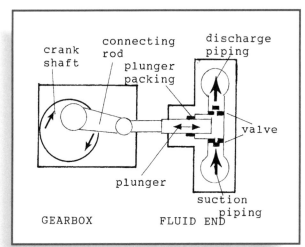

FIG. 11–2   Positive-displacement Pump

As the crankshaft is rotated by the V–belt, the connecting rod transmits reciprocating motion to the plunger. In Figure 10–2 the plunger is midway in the discharge stroke with the suction valve closed and the discharge valve open.

Positive–displacement pumps are used for high–pressure, moderate to low volume applications, such as water injection. Since they are more complex than centrifugal pumps, they require more maintenance. Typical configurations are simplexes (one plunger), triplexes (three plungers), and quintuplexes (five plungers). The stroke cycles of the multiple plungers are staggered to balance the load.

Compressors

Compressors do the same for gas streams as pumps do for liquids—creating a pressure differential that moves the gas through the system.

Most oilfield compressors are large positive–displacement, reciprocating machines used to inject low–pressure associated gas into high–pressure pipelines. Large piston size is necessary because compression ratios are as high as 10:1. That is, the gas is compressed to $1/10$ of its initial volume. These machines often have multiple compression stages, and are "integrated," with the gas–fired engine and compressor in a single package.

Centrifugal compressors are infrequently used in production operations. Pipeline boosting is the primary application for these extremely high–speed, high–volume machines.

Tanks

Tanks are large vessels used for volume storage of oil or water or to provide retention time for gravity segregation. Although storage tanks are not constructed to contain pressure, a few ounces of gas pressure is normally maintained to exclude air.

Pipe and fittings

Line pipe provides the conduits for transporting produced fluids to and through the facility. Pipe materials include:

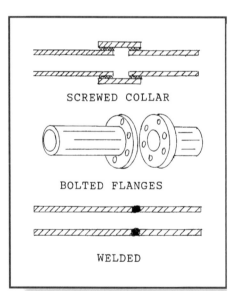

SCREWED COLLAR

BOLTED FLANGES

WELDED

- **Steel** is the most commonly used line pipe material. Its joints are usually welded or screwed together.
- **Plastic** pipe is used for low–pressure corrosive applications.
- For corrosive conditions at higher pressures, **fiberglass–epoxy** pipe is available, but very expensive.

FIG. 11–3 Pipe Couplings

Pipe fittings include:

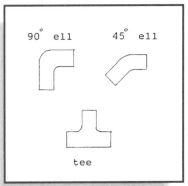

- *Collar, flange, and welded couplings* (FIGURE 11–3).

- *Elbows (ells)* to change pipe direction and *tees* to join three pieces of pipe (FIGURE 11–4).

FIG. 11–4   Pipe Fittings

Valves

Several types of valves are used in surface facilities. Their function is either blocking (open or closed) or control (throttling). Some of the common valves are

- *Butterfly (wafer) valves* are inexpensive block valves used for low–pressure, non–critical service; they consist of a steel wafer inside the pipe that rotates 90° to either block or allow flow

- *Ball valves* are used in higher–pressure, more demanding situations; a sphere with a conduit through it allows flow, or if rotated 90°, seals against flexible rings to block flow

- *Check valves* have a hinged flapper that swings open to allow flow in one direction but swings shut and blocks flow in the other direction

- *Globe valves* are the most commonly used control valves; they raise or lower a plug to increase or reduce flow

Meters

Metering functions

1. *Custody transfer (sale of oil and gas)* is the most demanding metering function, requiring extreme accuracy.

2. *Routine data gathering* requires less accurate measurement. Examples are the metering of each well's oil, gas and water producing rates during periodic well tests and the metering of salt water disposal volumes.

3. *Process monitoring and control* meters sense flow rates and pressures in vessels and piping, then alert the operator or automatically take the needed action.

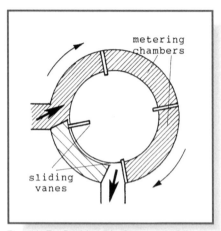

FIG. 11–5 Positive Displacement Oil Meter

Oil Meters

Rotary positive–displacement meters are used for most oil custody transfers.

The PD meter in Figure 11–5 has sliding vanes that form very precise metering chambers.

Other rotating vane or paddle type meters are less accurate, but entirely satisfactory for other than custody transfer measurements.

Turbine meters are finding increasing application, particularly in high–volume facilities. The fluid flow impinges on the turbine blades, accelerating the impeller's rotation in proportion to the rate of flow.

Metering chambers in pressure vessels are often used for well test measurement. Liquid fills the chamber until it reaches a level that triggers the opening of a dump valve.

Gas meters

Most field gas measurement is by orifice meter (FIGURE 11–6), but turbine–type meters are sometimes used in high volume installations. Orifice metering operations consist of the following steps:

1. Turbulence in the gas stream is reduced by flowing it through a straight, smooth, meter tube

2. The gas stream funnels down to pass through a limited–diameter orifice drilled through a steel orifice plate placed in the meter run

3. Taps on either side of the plate are used to sense the static pressure upstream of the plate and the differential pressure across the plate

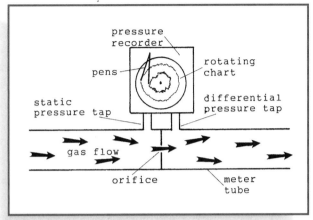

FIG. 11–6   Orifice Meter Facility

4. The two pressures are recorded either by pens on a rotating paper chart or electronically by a microprocessor

5. The volume of gas passing through the plate is then calculated based on the

    •   Static and differential pressure

    •   Periodic gas density tests

    •   Orifice and meter run size

Gas moving through a piping system constantly expands and contracts from the normal variations in pressure and temperature. It is therefore not possible to measure gas flow as accurately as oil flow with a PD meter. While at best, gas measurement by orifice meter is somewhat crude, in practice it frequently is highly inaccurate due to poor maintenance of the facility. Some of the common problems are:

•   Wearing of the sharp edge of the orifice results in erroneously low throughput readings

•   In field operations it's not unusual to have upsets where a slug of liquid passes through the gas system. The impact of the slug hitting the orifice plate often bows the plate—resulting again in low throughput readings

•   Buildup of hydrocarbon solids ahead of the plate reduces throughput readings.

Note that in all three examples above, the measurement error is in the purchaser's favor, working against the producer. In the U.S. particularly, but also in many places around the world, custody–transfer meter runs are owned and maintained by the purchaser. It's clear, however, that the producer must actively oversee meter maintenance to protect his interests.

Equipment Systems

The primary activity in surface processing facilities is separating the produced fluids into streams of oil, gas, and water for sale and disposal. This separation is accomplished largely by gravity segregation, drawing the lighter gas drawn off the top of vessels, the heavier water off the bottom, with only the oil remaining. High–gravity, low–viscosity crude oil normally separates easily from water, but lower gravity, higher viscosity crudes often form emulsions that complicate separation.

Flow lines

The produced fluids flow from the wellhead to central treating facilities through a flowline. The fluids are driven down the line and through the treating facilities by wellhead pressure, either natural or generated by artificial lift equipment.

On land, it's not unusual for as many as 50 wells to be served by a single tank battery (central treating facility). Flowlines are typically made of small diameter steel pipe, either welded or screwed. They may either be buried or laid on the surface. Coiled polyethylene is sometimes used in low–pressure applications. The lines radiate inward from all directions and some may be over a mile (5280') in length.

Fields in shallow, protected waters that are developed with dispersed vertical wells and a central facilities platform use flowlines similar to those on land. They may be laid on the seafloor or buried to avoid being hooked by dragging ship anchors.

On deepwater offshore platforms that support multiple trees for directional wells, the flow lines are only a few feet long—just enough to reach from the trees on the cellar deck to the valve manifold on the top deck.

Piping manifolds

Whether on land or offshore, all flowlines come together at the "manifold"—the first segment of the central processing train. This piping and valving station switches the flow from any well or group of wells into any one of multiple "headers" (FIGURE 11–7).

The typical system has a production header leading to the production treating system and a test header leading to the test system. One well at a time is switched into the test header while all the other wells are switched through the production header.

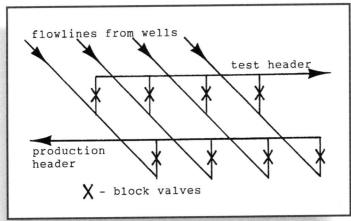

flowlines from wells

test header

production
header

X - block valves

FIG. 11–7 Piping Manifold

Gas separators

The first pressure vessel the fluid enters at the central facility is the gas separator, usually called "the separator." It separates the free gas from the liquid. Since the gas occupies many times the liquid volume, this sharply reduces the size of downstream facilities needed. Most separators are two–phase, separating the gas phase from the liquid phase, which includes both water and oil (FIGURE 11–8). Three–phase separators that divide the production into three streams—oil, water and gas—are sometimes used with high–gravity oils that readily separate from the water.

Vertical separators are used for low–volume operations, while horizontal separators are typically used in high–volume, high–GOR applications.

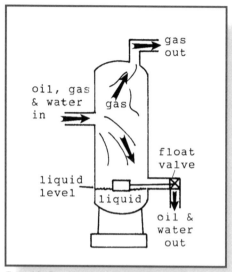

FIG. 11–8 Vertical Two-phase Gas Separator

The sequence of events in a gas separator is as follows:

1. The oil, gas, and water enter midway up the vessel.

2. Since the gas is lighter than the liquid, it rises upward and exits through the top of the vessel.

3. The liquid drops to the bottom of the vessel and accumulates. When the liquid level rises to a pre–determined point, the dump valve opens, dumping liquid. The valve closes when the liquid level reaches the bottom setting.

Free–water knockouts

Free–water knockouts (FWKs) are installed downstream from the separator when the wells produce a large volume of free water (water not tied up in an emulsion).

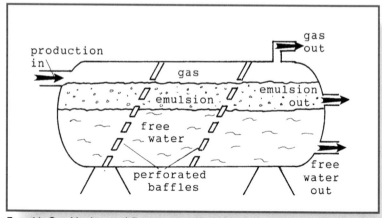

FIG. 11–9 Horizontal Free-water Knockout

Knockouts are large, usually horizontal, pressure vessels providing adequate retention time for free water to drop out of the emulsion. The free water is drawn out of the base of the vessel, the emulsion outlet is midway up the vessel, and any remaining gas is drawn off the top (FIGURE 11–9).

Emulsion–breaking facilities

Emulsions are mixtures of two insoluble liquids with small quantities of one of the liquids in small, discrete droplets dispersed throughout the other, continuous liquid. Oilfield emulsions are usually water–in–oil (the water is dispersed and the oil is continuous) and are often quite stable.

Produced water is considered a contaminant of crude oils, so purchasers accept only traces of S&W (sediment and water). Breaking emulsions to remove S&W is therefore a major production problem, particularly onshore where crudes are typically more viscous, and therefore more emulsion–prone. Emulsion problems increase in cold weather because the oil viscosity increases, tightening the emulsions.

Oilfield emulsions are broken in the surface facilities by several processes:

- *Applying heat.* Heat reduces oil viscosity and coalesces small water droplets into bigger ones. Both effects help the water to settle to the bottom of the treating vessel.

- *Injecting detergent–type chemicals.* Which weaken the surface tension between the oil and water, helping water coalescence.

- *Applying an electric potential.* When applied across the emulsion electrically charges the water droplets, causing them to coalesce.

- *Retention.* Wash tanks, or gunbarrels, are large tanks used to provide a great deal of quiet retention time for the water to settle out of the crude and be drawn off the bottom.

  Produced fluids enter most wash tanks through flumes designed to reduce turbulence in the tank. The clean oil spills over to stock tanks through an outlet near the top of the wash tank, the free water is drawn off the bottom. The incoming fluid may or may not have been heated, depending on how tight the emulsion is (FIGURE 11–10).

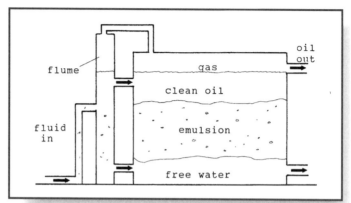

FIG. 11–10 Wash Tank with Flume

- **Heat.** Indirect heaters are used to heat emulsions, often upstream of a wash tank. The emulsion is circulated through heat exchanger coils immersed in a water–bath heated by a gas fueled fire–tube (FIGURE 11–11). Indirect heaters avoid the coking and fire–tube burnout problems encountered with direct heaters, which have fire tubes immersed directly in the emulsion.

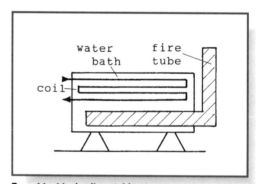

FIG. 11–11 Indirect Heater

Other treatment vessels include the following:

- **Heater treaters** are essentially three–phase separators with fire tubes. This combination provides a low–cost treating system for low–volume, marginal producing fields. Heater treaters can be either vertical or horizontal vessels.

- *Electrostatic treaters* have high voltage, alternating current electrical grids mounted inside a horizontal pressure vessel. A fire–tube is often installed in the other end of the vessel.

- *Desalters* remove salt granules from crude by washing the crude with fresh water. Excessive salt in crude is not common, with most problems occurring in the Middle East.

- *Slug catchers* are extremely high volume, two–phase gas separators designed specifically to handle the high–volume alternating slugs of gas and liquid that arrive at the shore facilities from two–phase pipelines.

  To handle this volume of fluid, a single vessel would have a diameter so large that it could not be built strong enough to hold the pressure. Slug catchers therefore use multiple interconnected pipe laterals to get the necessary volume. These installations spread over broad areas and are often partially buried.

Automatic custody transfer (ACT)

Most crude oil is sold automatically through Automatic Custody Transfer (ACT) units as follows:

1. When the oil level in the run tank rises and triggers a high–level switch, the ACT pump pulls oil from the bottom of the run tank.

2. An electrical capacitance probe inserted in the crude stream monitors the S&W content of the liquid being pumped to detect "bad oil". If S&W contamination rises above the purchaser's maximum (often 1%), valves are automatically switched to divert the crude stream back through the treating facilities.

3. If S&W remains acceptable, the crude volume is measured by an extremely accurate positive–displacement meter that automatically adjusts the recorded volume for temperature expansion or contraction. The meter is equipped with a printer than automatically prints the volume sold on the "run ticket" that verifies the sale.

4. Every few seconds, an automatic sampler extracts a small quantity of the passing crude stream and stores it in a sealed sample container.

5. Custody transfer is now complete and the crude enters the purchaser's pipeline, marine tanker, or other facility.

6. Periodically, often once a week, representatives of the producer and the purchaser meet at the ACT unit to process the accumulated sample.

   The sample is thoroughly mixed and a test–tube full is "shaken–out" in a centrifuge to determine the percent of S&W it contains. The S&W volume will be deducted before payment is made. Another portion of the sample is tested with a hydrometer to determine the crude's API gravity.

   The S&W and gravity data, witnessed by both parties, is entered on the run ticket, which already has the metered volume printed on it. The run ticket is then the basis for payment.

7. Every few months, the meter is "proved" by installing a highly calibrated check meter in series with it. Crude is pumped through both meters and the quantities they each record are then used to calculate the meter factor that is used to adjust the metered volumes until the next proving.

To review, the following data is required to transfer custody of crude oil

- volume
- temperature
- S&W content
- API gravity

Vapor recovery units

Vapor recovery units capture the low pressure gas that continues to vaporize out of crude oil as it passes through gun barrels, stock tanks, etc. The vapors are compressed into the sales gas stream. Vapor recovery saves valuable gas that would otherwise be flared and also reduces air pollution.

Gas well field processing

Gas wells, particularly dry–gas wells, have high wellhead pressures—essentially reservoir pressure—because of the negligible weight of the gas column in the tubing. Since wellhead pressure is typically much greater than pipeline pressure, a significant pressure–drop is taken across the choke on the well's Christmas tree.

Effects of choking. Through its contact with connate water in the reservoir, produced gas is fully saturated with water vapor. For example, gas produced at 120°F and 4000 psi. contains about 60 pounds of water per MCF. When the well is produced and pressure is reduced across the wellhead choke, the gas expands and draws heat from the surroundings (Joule–Thompson effect). This cooling reduces the quantity of water the gas can hold in solution, so water drops out.

Cooling across the choke also reduces the quantity of natural gas liquids (NGL) that can be held in solution, causing hydrocarbon liquids to condense out along with the water. This is called "condensate" when produced from rich gas streams that produce large quantities of liquid. In leaner streams, it is called "drip gasoline".

If the cooling is severe enough, ice can form in the choke and plug off production. Even if the temperature stays above freezing, gas hydrates can form below the choke and plug the pipe. Hydrates are compounds formed from gas and water that look like snow or ice crystals, but form at higher temperatures than ice.

Avoiding ice and hydrates. High pressure gas wells typically have a high pressure line connecting the tree with the choke, which is mounted on an indirect heater some distance away. When gas is produced, the heater's hot–water bath prevents temperatures in the coils downstream of the choke from dropping low enough for hydrates or ice to form.

The water and hydrocarbons that drop out must be removed so that they don't freeze further down the pipeline. The indirect heater and a separator are therefore often combined in a single vessel.

When the problem is not severe, or occurs only sporadically, it may be more cost–effective to inject liquid inhibitors into the gas stream. Either methanol or glycol are effective in lowering the freezing and hydrate–forming temperatures.

Dehydration. Dehydration is the permanent solution to ice and hydrate problems. It also prevents water from condensing in gas pipelines, interfering with flow. Full dehydration may not be necessary for relatively short in–field gathering systems, but is essential before gas is put into long–distance pipelines where considerable cooling takes place.

Two types of dehydration are in common use:

1. *Glycol dehydration units* bubble the gas through liquid glycol in a contactor tower. The glycol absorbs the water out of the gas and is then regenerated by distillation. Because of their compactness, glycol units are favored for offshore facilities.

2. *Solid desiccant units* percolate the gas upward through a tower filled with dry desiccant. While one tower is in use, the water–saturated desiccant in a the second tower is being regenerated with hot air.

Pipelines

Pipelines provide surface transport for both gas and liquids. They are constructed by welding together joints of large–diameter steel pipe. Most pipelines are externally coated and wrapped for protection from corrosion and abrasion. Onshore pipelines are buried. Shallow–water offshore lines are usually buried for protection, but deeper–water lines may not be buried. Single–phase pipelines transport either gas or liquid, but not both. Two–phase pipelines transport both liquid and gas. In two–phase lines, liquid builds up in the low spots (FIGURE 11–12) and then starts moving as a slug (filling the entire line). The alternating high–volume slugs of gas and liquid emerging at the end of the line make severe demands on the separation equipment.

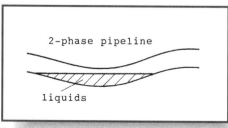

2-phase pipeline

liquids

FIG. 11–12 Liquids Accumulating in Pipeline Low Spots

Pipelines are used both in gathering systems—moving produced fluids between facilities within the field—and for final transport of fluids away from the field.

Rubber swabs or spheres called pigs are periodically run through pipelines to sweep out accumulated water and debris. The pigs fit tightly, forming a seal against the inside of the line so that fluid pressure propels them. They are inserted and removed from the pressurized line through a pig launcher at the beginning of the line and a pig trap at the end of the line.

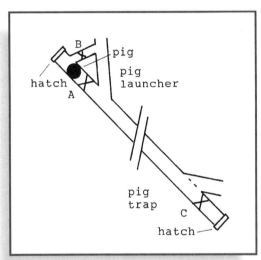

FIG. **11–13** Pipeline Pigging System

Referring to Figure 11–13, the procedure used to pig a line is:

1. The pig is loaded by closing valves A and B, bleeding off launcher pressure, opening the hatch, inserting the pig and closing the hatch

2. The pig is launched by opening valve A to provide access to the main line and then opening valve B to blow the pig out of the launcher; valve C is also open

3. The pig traverses the pipeline and lodges in the pig trap

4. The pig is recovered by closing valve C, bleeding off pressure, opening the hatch and removing the pig

Facility Layout

Onshore

Onshore wells are normally drilled vertically, so their wellheads are scattered over the field's surface. Flowlines radiate inward, connecting each well to a manifold at the central tank battery. Wellhead pressure, either naturally occurring or generated by artificial lift equipment, moves the produced fluids through the flowlines and tank battery (FIGURE 11–14).

- The manifold valves are opened and closed to send one well at a time through the well test system, where the well's production rate of oil, gas, and water is measured. Production of all the other wells is directed through the normal production system.

- The oil, gas, and water are separated by gravity segregation. Emulsion–breaking measures are usually required.

- The oil is metered and sold, usually through an ACT, into the purchaser's pipeline or tank truck.

- The gas is sold by metering its volume through an orifice meter and its composition is periodically checked to determine the content of ethane and heavier hydrocarbons. The unit price of the gas is then adjusted to reflect the additional heat content.

  This rich (high content of heavier hydrocarbon vapor) and wet (high water vapor content) gas cannot be transported very far before it cools and condenses liquid out in the pipeline or forms gas hydrates. From the tank battery, the gas is therefore pipelined to a nearby natural gas processing plant where the hydrocarbon liquids and water are removed and the residue gas is compressed into the sales pipeline.

- The produced water, which is usually too salty to be dumped into surface waters, must be disposed of into subsurface formations. If there is a waterflood in the vicinity, the water can be put to valuable use.

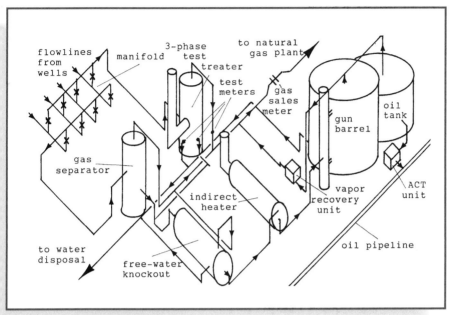

FIG. 11–14 Onshore Tank Battery

Shallow waters

The development scheme in shallow water is identical to onshore. Individual jackets are set and the wells are drilled vertically. Long flow lines connect them to a central platform where processing takes place.

Deeper water, moderate sea states

In deeper water, individual well jackets become too costly, so multiple directional wells are drilled from a single well platform.

Several well platforms are connected by three–phase (oil, water, and gas) pipelines to a facilities platform where processing takes place. If the field's location is far offshore, a quarters platform is also installed.

Wells, facilities and quarters are kept on separate platforms if possible, although they may be tied together by walkways. This is a safety measure reflecting that fired processing facilities have the highest risk of all offshore facilities. It is therefore desirable to separate the wells and the personnel from this risk as much as is possible.

The following operations take place on the facilities platform (FIGURE 11–15):

- The oil, gas, and water are separated. Since offshore crudes generally have relatively high API gravity, emulsion problems are usually not severe. Horizontal, three–phase separators, therefore, work quite well.

- All traces of oil are removed from the produced water, which is dumped overboard into the sea.

- The produced gas is dehydrated to avoid hydrates, and compressed into a pipeline to shore. Liquid glycol dehydration systems are normally used offshore.

- The crude oil is pumped into a pipeline to shore. If the same pipeline is used for both gas and oil, two–phase flow will be present.

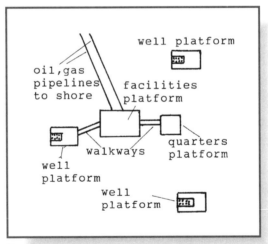

FIG. 11–15 Offshore Platform Layout

Deep water, severe sea states

As the water gets deeper and platform jackets get more expensive, operators must forego the luxury of separate well, facility, and quarters platforms. Instead, the three functions are combined into a single platform, which also functions as a drilling platform in the early stages.

The single platform is inherently less safe than separate platforms, but there is no other possibility in places like the North Sea, where a single platform can cost well over a billion U.S. dollars.

Subsurface completions

With subsurface completions, the wells are remote from the facilities but connected by individual flowlines or by pipelines serving several wells. The surface facilities can be mounted on bottom–supported platforms or floaters such as tension–leg platforms or FPSOs.

NATURAL GAS 12

Natural gas is a highly desirable fuel because

- It burns completely without smoke or noxious combustion products. This makes it ideal for a host of applications as diverse as commercial food preparation and vehicular use.

- The burner systems for natural gas are simple, cheap and do not become fouled with carbon. This is a particular advantage in small–scale commercial processes.

- Natural gas has a higher hydrogen–to–carbon ratio than fuel oil or coal, resulting in 30% less carbon emissions than oil and 50% less carbon emissions than coal to yield the same quantity of energy. This environmental advantage encourages gas use for power generation and other large industrial applications.

Remote Gas Fields

Natural gas serves a relatively local market, while crude oil is a worldwide–traded commodity. This is because natural gas cannot be shipped around the world in conventional marine tankers. Crude oil found anywhere on earth can be economically brought to market. Remote gas fields, however, are usually uneconomic because transportation of the gas is not practical. The only options to commercialize a remote gas field are to:

- Bring the market to the field by, for example, constructing a fertilizer plant at the remote site; being a higher valued product, the fertilizer can then be economically shipped to market

- Install an LNG facility, but as discussed later in this chapter, there are problems with this

Field handling of natural gas

The natural gas produced in association with crude oil is recovered largely from the gas separators, with lesser amounts taken from other vessels in the tank battery. It typically is low pressure, saturated with water vapor and heavier hydrocarbons (NGLs), and is often contaminated with hydrogen sulfide and carbon dioxide.

Some of the gas is used at the facility to operate fired vessels, control systems, pumps, compressors, gas–lift systems, etc. Occasionally, some of it may be flared when there is an upset in the treating system. In the United States where mineral rights are generally privately owned, it's not unusual for a tank battery to serve wells with various ownership interests. To preserve equity, it's necessary to keep track not only of the gas volume each well produced but also how much it consumed in gas lift and other onsite requirements (FIGURE 12–1). In these circumstances, the measurement and accounting for the various gas streams can be extremely demanding. The requirement is that each on–site use be allocated fairly among the wells so that the revenue from gas sales can be equitably distributed to the various interests.

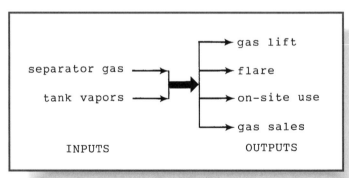

FIG. 12–1   Gas Streams in Field Facilities

The gas remaining after on–site use is then transported by pipeline a short distance to a natural gas processing plant where it is prepared for long–distance pipelining. Since the plants benefit from economies of scale, a single large plant often serves multiple fields in the general area. Onshore in the United States, the gas plants and field gathering systems are usually not owned by the producers. Instead, separate companies buy the "wet" gas from the producers as it exits the tank battery through the orifice meter run. Elsewhere, gas processing facilities are usually owned and operated by the producers.

Gas–well gas is recovered at much higher pressures than associated gas—wellhead pressures are typically several thousand psi—so there is less water and NGL in solution. This requires less severe processing.

Disposal of Surplus Associated Gas

Oil wells are typically produced at capacity because there is a ready market somewhere in the world for all the crude that can be produced. Gas produced with the crude, however, is essentially a byproduct and gets produced whether there is a ready market for it or not. It's not feasible to store the surplus—the huge pressure tanks that would be required are totally impractical. As a result, surplus associated gas is essentially valueless, and in the past was usually flared and burned. In recent years, flaring has been substantially reduced around the world by the injection of the surplus gas back into the reservoir. The rationale is that

- Flaring has been recognized as an environmental problem because methane and its combustion products are "greenhouse gases" that may contribute to global warming.

- It intuitively seems wasteful to throw away such a useful substance that may have considerable value in the future.

- The injected gas can enhance oil recovery from the reservoir.

In contrast to associated gas, gas–well gas is produced only when there is a market demand for it. This "pipeline proration" means that gas–well gas is essentially stored in the reservoir until someone wants to buy it. The wells are opened up when the pipeline inlet pressure drops and closed in when it rises. This avoids having to dispose of surplus gas by flaring or injection.

Natural Gas Processing

Gas processing plants extract heavier petroleum products and contaminants from the residue gas stream to prepare it for long–distance pipeline transmission. These include (FIGURE 12–2)

- *Natural gas liquids (NGLs)*, which include pentane, hexane and heavier gasoline–range molecules, are removed to prevent them from condensing in the pipeline and interfering with gas flow. Most NGL is sold to refineries as gasoline feedstock. The removal of NGLs therefore both benefits the pipelining operation and generates significant revenue.

- *Water* is removed so it doesn't condense in the pipeline and interfere with gas flow, form ice and hydrates that could plug the line, or contribute to internal corrosion.

- *Liquefied petroleum gas (LPG)*—propane and butane—is extracted both to prevent condensation in the pipeline and for sale as petrochemical feedstock, refinery feedstock, and bottled gas.

- *Ethane* is extracted for sale as petrochemical feedstock if there is a good market for it in the vicinity. If no market is available, it remains with the methane, increasing the heat value and price of the residue gas. Ethane is sufficiently volatile and will not condense in the pipeline.

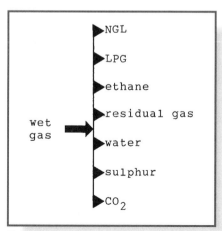

FIG. 12–2   Natural Gas Inputs and Outputs

- *Hydrogen sulfide gas (H₂S)* is a corrosive contaminant and must be removed. This yields elemental sulfur, which is sold to chemical plants.

- *Carbon dioxide (CO₂)* is a contaminant without heating value and is corrosive in the presence of water vapor. There is usually no market for the extracted CO_2, so it is flared to the atmosphere. In the western United States, however, considerable quantities of extracted CO_2 finds a market in miscible enhanced recovery projects.

The processes used in gas plants are as follows:

- *Compression* to pipeline pressure condenses most NGLs and some LPG.

- *Absorption* involves bubbling the gas stream upward through a contactor tower containing downward moving absorption oil. All of the LPG and some of the ethane are recovered by absorption.

- *Cryogenic (very low temperature)* processing is used for high ethane recovery. The low temperatures are obtaining by expanding high pressure gas.

- *Sweetening* removes hydrogen sulfide and carbon dioxide, called "acid gases", by reaction with chemicals such as monoethanolamine (MEA).

- *Dehydration* by glycol or solid desiccant removes the water vapor.

Natural gas pipelines

Welded steel pipelines are the dominant mode of transporting natural gas to market. Gas pipelines are typically large diameter (20 to 40 inches) and are operated at relatively high pressures (800 to 1000 psi). They are externally coated and wrapped to prevent corrosion and are usually buried for protection.

The pipeline infrastructure necessary to commercialize natural gas requires enormous investment, but yields decades of profits. Two examples of such investments that are now generating large returns are the transcontinental lines in the United States extending from the producing areas in the southwest to the major consuming areas in the northeast, and the Russian lines extending from western Siberia to western Europe. In contrast, Iran is second only to Russia in gas reserves, but produces insignificant quantities because there is no infrastructure in place.

LNG
(liquefied natural gas)

Liquefying methane into LNG generates a compact, higher valued form of energy that can be economically transported around the world by ship. This allows the development of gas fields that would otherwise be uneconomic because they are too remote from potential consumers to justify construction of a pipeline.

The purpose for changing methane from a gas to a liquid is entirely to facilitate its transportation. There is no market for the liquid as such, so it is regasified at the landing port and delivered to the consumer through conventional natural gas pipeline systems. In some cases, it is blended into a gas stream to supplement local supplies of natural gas. The problem with using LNG directly is that it must be stored at cryogenic (extremely cold) temperatures to prevent vaporization. Such heavily insulated storage is too bulky and expensive for most applications, but experimentation in direct use of LNG is ongoing. For example, its use in large trucks is being examined.

Upstream facilities

The first stage in developing an LNG project is to install the upstream facilities necessary to deliver gas to the liquefaction plant. This involves developing the gas field, installing a control system to regulate gas flow and, if offshore, constructing a pipeline to transport the gas to shore. Control of the flow rate to assure uninterrupted charge to the plant is critical, given the huge volumes of gas being handled. Some of the intensifying factors are the plant's 24–hour per day operation, the lack of surge–moderating storage facilities at the plant's suction, and, in some cases, two–phase flow in the pipeline delivering alternating slugs of gas and liquid.

Liquefaction plant

Although increased pressure and reduced temperature will liquefy petroleum gases, the extreme volatility of methane dictates that cryogenic cold (-260° F) be used to liquefy and transport LNG. The liquefaction plant therefore amounts to a very large refrigeration plant. The plants are built onshore and include marine loading facilities. In the future, floating plants may be used in some protected overwater situations.

Marine tankers

LNG is transported in specially constructed marine tankers and maintained in the liquid state by heavy insulation. Some vessels use spherical tanks set within the ship's hull while some use tanks integrated into the ship's hull and separated by membranes. Gas that vaporizes during the voyage is used either to fuel the vessel or is reliquefied by an onboard refrigeration unit.

There have not been any serious accidents involving LNG tankers. Nevertheless, it needs to be recognized that, given the extreme heat content and volatility of LNG, a tank–rupturing incident would cause an unprecedented conflagration.

The scheduling of tanker movements is quite demanding in view of

- The impracticability and therefore lack of significant LNG storage capacity between the plant outlet and the loading dock. Given the unstoppable nature of the product stream from the wells, through the pipeline, into the plant and out to the ships, the shipping operation must never be caught without a ship to receive product.

- The necessity to arrive at the destination in time to fit smoothly into the offloading operation at the end of the voyage. The product flow on the consumption end is as fully unstoppable as it was at the source. The ships must be offloaded, the LNG vaporized, the gas injected into the delivery line, and ultimately delivered to the customer without interruption or significant fluctuation.

Regasification terminal

The process of regasification is much less intense than liquefaction, but with the enormous quantities of energy involved, it must be handled carefully. Explosions have happened in this operation. As it was on the other end, controlling the flow rate of gas to the consumer is critical.

Economics

The economic justification of an LNG project faces several major hurdles:

- LNG plants are subject to strong economies of scale—small plants are simply not economic. Capital requirements are therefore huge, with billions of dollars required for a project. The sheer size of the capital required limits the number of potential sources for it.

- Most of the capital is invested in fixed assets located onshore in developing countries. This property could be vulnerable to local disturbances or nationalization. The concern is amplified by the long–term (15 to 20 years) nature of these projects.

- The LNG industry is small and poorly integrated. This means that the utility of each component of a project—the gas field, pipeline, liquefaction plant, shipping port, tankers, and regasification terminal—is dependent on the overall functioning of that particular project. If one link in the chain fails, it is difficult to develop alternative opportunities to realize value from the remaining links.

- The profitability of LNG projects is, at best, modest. There are enormous quantities of undeveloped gas around the world trying to squeeze into the very limited LNG market. This places purchasers in a very strong negotiating position, from which they trim suppliers profit margins to the vanishing point.

The huge capital cost to construct the infrastructure is what limits the size of the market for LNG. The cost of simply servicing the debt—not to mention a profit for the supplier—typically exceeds the price of the fuel oil alternative. This puts such a high floor under the price for LNG that the only consumers are those whose applications require the unique characteristics of natural gas and who absolutely have no alternative supply of it available. High costs have held LNG to a minor fraction of the world market for natural gas, and will continue to limit its growth in the future.

Supply and demand

The market for LNG is dominated by Japan, which consumes most of the world's supply. South Korea is the second largest consumer. The balance of demand is scattered among several European and Asian countries.

Asia supplies most of the world's LNG. Indonesia is the largest producer, followed by Malaysia, Australia, and Brunei. Algeria is the only significant producer outside of Asia.

GTL Processing
(Gas–to–Liquids Processing)

GTL at this writing is still in the research phase, not yet a commercial reality. However, it would be an oversight to not discuss the profound ramifications for the petroleum industry should it become established.

GTL processing converts methane to liquid petroleum products such as middle distillates. Variants of it have been around for some time. During World War II, Germany used a GTL process to manufacture gasoline to help run their war machine. Another process was developed in South Africa. Neither process is considered to be commercial in today's markets, but there is a great deal of work under way to refine the technology. Should this research be successful—and there is reason to be optimistic—GTL could become serious competition to LNG. It could even effect the market for conventional crude oil.

The promise and excitement surrounding GTL processing is that by converting gas to liquids, the product then enjoys the same ease of transportation that crude oil now has. This would suddenly make many remote, undeveloped gas fields around the world commercial.

GTL is much less capital intensive than LNG and does not require the economies of scale—small plants can be as profitable as large ones. This raises the intriguing possibility of employing small GTL plants on FPSOs and other remote production platforms. These installations typically have no pipeline connections. The produced crude is offloaded to shuttle tankers, but the associated gas must either be flared or injected into the reservoir. GTL processing would solve this problem by converting the gas into liquids that can be mixed with the crude for shipment to market.

A further bonus of GTL processing is that its origin in gaseous methane eliminates all sulfur and other contaminants from the resulting liquid products. Such purity is not obtainable in products derived from crude oil, and would significantly reduce environmental contamination and engine corrosion.

Given the potential advantages of GTL processing along with the enormous quantities of natural gas available around the world—both discovered and to be discovered—a great deal is riding on its development. There is even speculation that GTL could cause a shift away from crude oil as the world's fuel base.

REFINING AND PETROCHEMICALS 13

Crude oils are highly variable, with no two exactly alike. One crude may be bright green in color, sweet smelling, and as fluid as water. Another may be jet black and have an unpleasant sulfurous smell. Yet another may be pink and stack up in a pile when poured on the ground on a hot day. At the other extreme, gas condensate is colorless and will completely evaporate if left uncontained. The specific characteristics of each crude reflects its unique blend of sizes and types of hydrocarbon molecules, with thousands of different compounds in the mixture.

This extreme variability is why the material was named "crude". It is indeed a crude material, being messy, unstable, and dangerous to handle. Essentially unusable in its initial form, crude must be refined—broken down—into products with the specific characteristics to handle certain jobs well, e.g., gasoline to power cars, fuel oil to heat homes, etc. This is the role of refineries.

Feedstocks

The input to refineries is principally crude oil, including condensate, but also includes large quantities of NGL and LPG (mostly butane). Refineries are generally not designed for a specific crude oil, although such situations exist. Instead, they have considerable flexibility to shift feedstocks to adjust to market conditions.

231

To select the most profitable feedstock at any given time, process engineers employ computer models of the specific processing capabilities at the refinery. Using detailed crude analyses and current prices, sequential computer runs simulate the refinery's yield and resulting revenues for each candidate crude. Each crude's value is different for each refinery, and its value changes as product price changes. For example, a momentary tightness in the local gasoline market, resulting in a price rise for gasoline relative to other products, would temporarily advance the value of light (high gasoline yield) crudes relative to heavier crudes. These are some of the factors behind "posted prices" in the crude market.

Very little of the crude refiners run is produced by their own company. Instead, it is generally purchased on the open market. The industry has a very efficient market operating between production and refining. Producers can count on getting a good price for their crude anywhere in the world and refiners can count on being able to purchase feedstock at a good price. Crude is traded freely, often changing hands several times as it takes the most efficient route between the producing fields and the refinery.

Sulfur in crude

Refineries are classified as either "sweet," meaning they can handle only non–sulfurous crude, or "sour" meaning they can handle either sulfurous or non–sulfurous crude. The definition of sour crude is that it contains 2.5% or more sulfur, whereas sweet crude contains 0.5% or less sulfur. In between is "intermediate" crude. The sulfur in crude is not elemental; rather it is tied up in hydrocarbon molecules. This complicates its removal.

Refineries handling sour crude must have extensive internal corrosion protection such as stainless steel trim on valves, plastic–coated lines and tanks, etc. To compensate for these costs, sour crudes are priced lower than sweet crudes.

Crude Oil Transportation

Crude oil can be transported from the producing field to the refinery in a variety of conveyances. For onshore production, tank trucks, railroad tank cars, and barges on inland waterways are used to some extent, but pipelines are the dominant mode of transportation. Pipelines are also used to bring production ashore from offshore fields located reasonably near to land.

Fields in very deep water—often employing subsea completions and FPSOs—typically use shuttle tankers to move the crude from the storage tanks in the FPSO hulls to shore. For most export operations around the world, marine tankers are the only choice available. The bulk of the enormous Middle East production, for example, is shipped to all parts of the world in tankers.

Crude oil pipelines have established an excellent environmental record. They are robust systems that seldom fail. When they do fail, the pumps are immediately shut down, holding escaped crude to a minimum.

Marine tankers do not have quite as good an environmental record as pipelines. One problem is simply the huge number of tankers that are on the high seas at any given moment. With an operation that large, some accidents are bound to happen. The most serious incident was the grounding of the EXXON Valdez, a very large crude carrier (VLCC), in Alaskan coastal waters. VLCCs, which haul much of the world's crude, are essentially huge floating crude tanks that are extremely difficult to stop or maneuver. When they do hit something, the quantity of crude available for release is staggering. To minimize risk, VLCCs are routed as far as possible outside of heavy shipping lanes. Additional transhipment terminals are being constructed around the world for VLCCs to unload out to sea; then pipelines or small shuttle tankers handle the final shipment leg into congested ports. Also, many VLCCs are now being built with double hulls, which provide some protection against impact.

What Refineries Do

Crude oil cannot be used as a fuel, itself, because it contains such a broad spectrum of molecular size. To be practical as a fuel, a material must have known burning characteristics so that it can be matched with an appropriate burner system. For example, butane requires the simplest of burner systems and is consumed completely without residue. The heavy, black, viscous, residual fuel oil, however, must be heated and blown into the firebox in atomized form to burn at all effectively. It produces considerable coke and ash that must be handled. Butane and residual fuel oil are two of the range of products produced by refineries to meet the specifications of various burner systems.

Taking cuts out of crude

Smaller molecules in a hydrocarbon increase volatility and reduce viscosity and color. Conversely, larger molecules reduce volatility and increase viscosity and color. Since volatility and viscosity are the important characteristics in designing burner systems, it follows that the principal difference between the various hydrocarbon products is the size of their molecules.

The first step in a refinery is to split crude into several fractions or "cuts"; each cut concentrating a particular range of molecular sizes. For example, one of the cuts might be light gases described as "butanes and lighter"; a heavier cut might be kerosene. The cuts are made with a distillation process that employs selective vaporization and recondensation.

Altering molecules. The highest valued petroleum products are the "white products", which include gasoline, diesel, jet fuel, and heating oil. Production of these "middle of the barrel" products, particularly gasoline, is maximized by converting as much as possible of the lesser–valued gases and the heavier "bottom of the barrel" materials to white products. This is done by altering the molecules as follows:

- Polymerization is used to combine the small gas molecules into larger, middle of the barrel, molecules

- Cracking is used to break up the large, bottom of the barrel, molecules into smaller ones

- Isomerization alters the arrangement of atoms in a molecule without adding or removing any the atoms

The tool kit. The three tools that refiners have available are:

- Heat to excite the molecules

- Pressure to retard vaporization

- Catalysts to assist the reaction but not participate in it

Processes

Refineries vary widely in complexity. The simplest are little more than atmospheric distillation towers. Complexity increases when more intense processes are added to drive the light and heavy ends further toward the middle of the barrel. Listed below are some of the processes employed in a relatively complex refinery.

Atmospheric distillation. To understand how the atmospheric distillation tower works, its necessary first to understand the distillation process. Figure 13–1 shows tap water being boiled (vaporized), then cooled so that it recondenses. This process creates distilled water, which is purer than the original water because only the H_2O molecules vaporized and left the vessel. The various non–volatile impurities in the tap water remain behind, building up in concentration, and eventually precipitating out as a solid residue when the vessel boils dry. This process functions then to separate the water from the impurities.

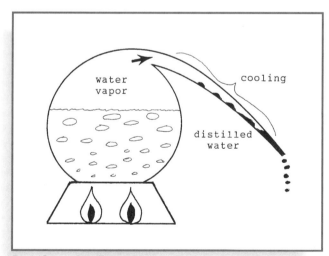

FIG. 13–1 Distilling Water

In the above example, water is boiled at 212° F (at sea level). The 212 ° F is called water's *boiling point* and is specific to the water molecule. Each of the hydrocarbon molecules has its specific boiling point, with the smaller the molecule, the lower its boiling point. This is the key to cutting crudes in a distillation tower.

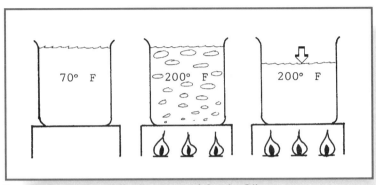

FIG. 13–2   Boiling Off a Fraction of Crude Oil

Figure 13–2 shows a beaker of crude oil on a Bunsen burner at room temperature. The burner is then ignited and the crude heated up to and held at 200° F. The crude initially boils vigorously, but the boiling eventually slows and finally stops. In the process, about a quarter of the crude boiled off. If the vapors had been cooled and condensed, the resulting product would be a concentrated solution of compounds with boiling points less than 200° F. The remaining solution would then be a concentration of those compounds with a boiling point greater than 200° F. Increasing the temperature restarts the boiling and a higher temperature cut is taken. In this way, crude oil is separated into fractions of similar molecular size.

Atmospheric distillation is the initial process in a refinery. Its fractionating tower—a large diameter vertical vessel with multiple internal contactor trays—is the largest in the refinery. The crude is heated, vaporizing most of it, then injected into the base of the tower. Vapors rising through ports in the trays come into contact with condensed liquids working downward through the column. This interchange continually vaporizes some liquid and condenses some gas, establishing an equilibrium throughout the column with heavier material toward the base and lighter material toward the top. Product cuts are continuously removed from the tower through side draws spaced along its vertical length. The location of each side draw is designed to tap a specific boiling point range of material. For example, in Figure 13–3 the light–gas oil cut has a boiling point range from 450° to 650° F.

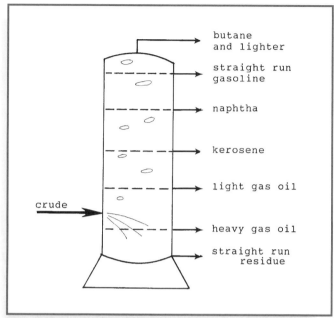

FIG. 13–3 Distillation Tower

Vacuum flashing

Feedstock: Residue from distillation

Purpose: Extract additional lighter material

Mechanism: Residue is heated and a vacuum pulled to induce boiling

Yield: Gas oils and residual

Flashing extends the distillation process to the bottom of the barrel. If done with heat alone, the high temperature required would induce cracking into lesser valued gases. By utilizing vacuum, lower temperatures are adequate and cracking is avoided.

Thermal cracking

Feedstock: Residual from flasher

Purpose: Split large molecules into gasoline–range molecules

Mechanism: Heat (920°–1020° F)

Yield: Full range of products including coke

Cat cracking

Feedstock: Heavy gas oils from distillation and flasher

Purpose: Convert heavy cuts to gasoline

Mechanism: Heat plus catalyst (beads or powder)

Yield: Full range from methane through residual

Cracking forms some olefins (FIGURE 13-4)—useful molecules that are deficient in carbon, e.g., ethylene (C_2), propylene (C_3) and Butylene (C_4). Olefins are manmade—not occurring in nature.

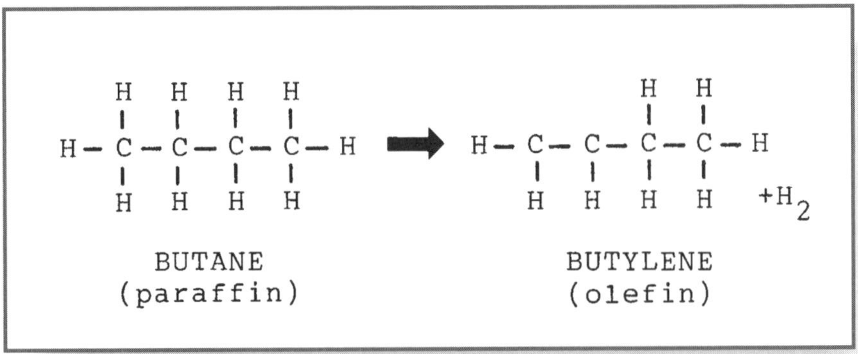

FIG. 13-4   Butylene Derived from Normal Butane

Hydrocracking

Feedstock: Gas oil from distillation

Purpose: Make gasoline out of gas oils

Mechanism: Cat cracking with added hydrogen

Yield: High quality gasoline components and no residue

Produces highest value products of any process. Useful in the seasonal refinery shift from fuel oil to gasoline production.

Gas plant

Feedstock: Gas streams from distillation and other processes

Purpose: Separate the gases for particular uses

Mechanism: Distillation under pressure, absorption

Yield: Methane, ethane, propane, butane, isobutane (FIGURE 13–5)

The products are all saturated (having a full complement of hydrogen atoms). Olephinic gases from crackers are separated similarly, but in a different gas plant.

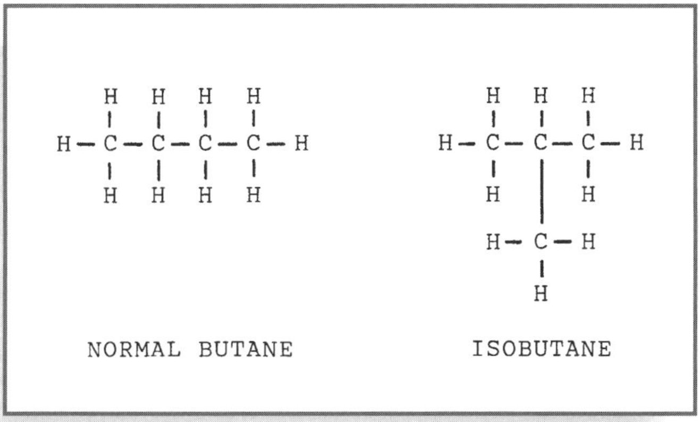

FIG. 13–5   Isobutane, an Isomer of Normal Butane

Alkylation

Feedstock: Olefins (propylene and butylene) from cracker plus isobutane from gas plant

Purpose: Create heavier, gasoline range molecules from lighter ends

Mechanism: Polymerization by pressure and cooling in presence of a catalyst (sulfuric or hydrofluoric acid)

Yield: Alkylate and gases

Catalytic reforming

Feedstock: Naphtha from distillation

Purpose: Reform paraffinic, gasoline–range molecules into aromatic, gasoline–range molecules, which have higher octane numbers (FIGURE 13–6)

Mechanism: Use of an exotic catalyst (alumina, silica and platinum) plus heat and pressure.

Yield: Gasoline–range aromatics plus gases

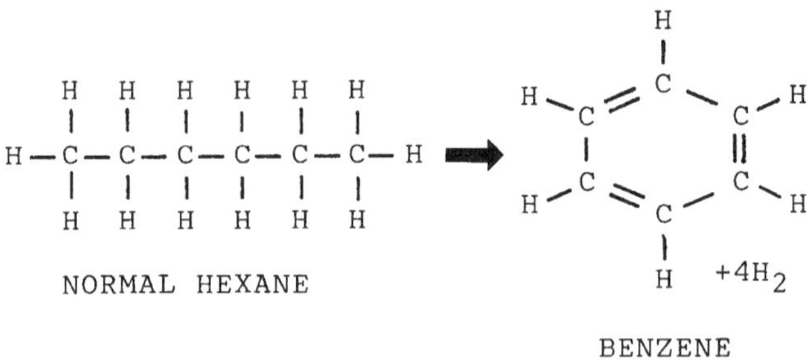

FIG. 13–6   Converting Paraffins to Aromatics

Blending

Blending operations are the final step, combining the various outputs of all the processes into the final products that are sold. The blending operation has a major impact on managing the refinery. It determines what feedstocks and run volumes are needed so each of the various processes will yield the required volumes and specifications of products.

The concerns in gasoline blending are with the vapor pressure and octane number of the various grades of gasoline produced. Vapor pressure specifications vary seasonally, with a more volatile gasoline needed in cold weather to assure ready engine starting. The pressure is reduced in warm weather to prevent vapor–locking the engine. It's also necessary to vary vapor pressure depending on the elevation—hence atmospheric pressure— of the location where its to be consumed. To prevent vapor–locking, gasoline blended for a high mountain area must be less volatile than that blended for a coastal area.

It is critical for gasoline octane numbers to be high enough to prevent engine knocking. Engines are designed for the spark plug to ignite the air–fuel mixture when the piston reaches the top of its stroke, which is the point of maximum air–fuel compression in the piston's cycle. If the gasoline's octane number is too low, it will spontaneously ignite before the piston reaches the top. The pulse, or knock, generated by this out–of–sync firing is in the opposite direction to the movement of the rising piston. This causes loss of power and can damage the engine.

In the past, octane ratings in gasoline were reached by adding tetraethyl lead. With the use of lead now banned in most countries, refineries now get their octane enhancement by more intensive processing, such as alkylation and reforming, and by adding alcohols and oxygenates. Three of the alcohols used are methanol, ethanol, and TBA (tertiary butyl alcohol). MTBE (methyl tertiary butyl ether) is an oxygenate used.

Government agencies proscribe various anti–pollution specifications for gasoline. For example, vapor pressure may be restricted to limit vapors escaping to the atmosphere while filling auto tanks. Similarly, the addition of alcohols or oxygenates may be required because their oxygen content reduces tailpipe emissions of ozone and other pollutants.

Diesel fuel and furnace oil (#2 fuel oil) are blended from the light gas oil streams. The two products are nearly identical and often used interchangeably. The critical characteristic for diesel fuels is to have a relatively low self–ignition temperature. For furnace oil, the critical characteristics are the safety issue of "flash point", involving volatility and flammability, and the "pour point" which is the ability to flow at low temperatures.

The heavy, black, bottom–of–the–barrel residual remaining after processing can be sold as asphalt or blended with lighter materials and sold as "resid" (residual fuel oil). Dense and viscous, resid is used as fuel for large industrial processes where its practical to install the complex preheating and atomizing burner systems needed.

Petrochemicals

Petrochemicals are non–fuel compounds derived from crude oil and natural gas. The industry has been built on the extreme reactivity of the carbon atom. Carbon's four valence bonds give it a strong propensity to bond with other atoms to form stable compounds. Hydrogen is the most commonly bonded atom, followed by oxygen and nitrogen. Sulfur, chlorine, and others are also present in some compounds. Adding to this diversity, the capacity to form multiple isomers of the same group of atoms extends the count of organic (carbon) compounds into the many thousands. In fact, organic compounds represent 95% of the million plus compounds on earth.

The feedstocks for petrochemical plants are provided largely by refineries and include gases, naphtha, kerosene, and light gas oil. Natural gas processing plants are also a source of feedstock, providing natural gas, ethane and LPG.

Petrochemical plants are typically built adjacent to refineries to be near their source of feedstock. Also, there often are byproduct streams created in the plants that can best be utilized by returning them to the refinery for processing. In fact, the boundary between the two types of plant often becomes blurred as progressive integration of both fluid streams and operating personnel takes place for increased efficiency. As a result, several very large refining/petrochemical complexes have developed around the world.

Petrochemical processes are, in general, much smaller than refinery processes, but there are exceptions like very large ethylene plants. The same tools are used as in refining—heat, pressure and catalysts—but, unlike refineries, inorganic chemicals are used in some of the reactions.

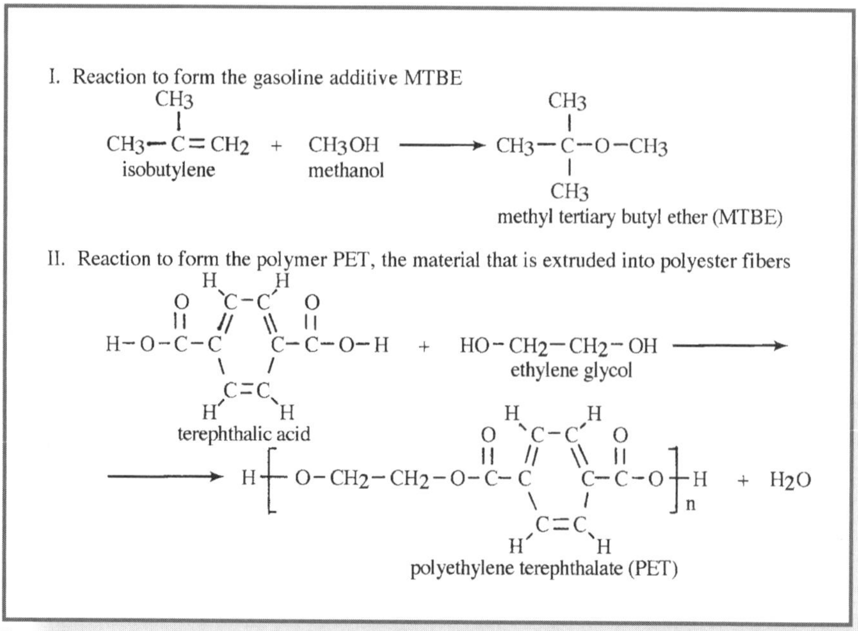

I. Reaction to form the gasoline additive MTBE

$$CH_3-\underset{\underset{isobutylene}{}}{\overset{\overset{CH_3}{|}}{C}}=CH_2 \;+\; \underset{methanol}{CH_3OH} \longrightarrow CH_3-\underset{\underset{CH_3}{|}}{\overset{\overset{CH_3}{|}}{C}}-O-CH_3$$

methyl tertiary butyl ether (MTBE)

II. Reaction to form the polymer PET, the material that is extruded into polyester fibers

terephthalic acid $+$ $HO-CH_2-CH_2-OH$ ethylene glycol $\longrightarrow$

polyethylene terephthalate (PET) $+\; H_2O$

FIG. 13–7 Petrochemical Examples and Uses

The examples of petrochemicals and their uses in Figure 13–7 give an indication of the extreme value of these molecules to our well–being. It's not hard to visualize a future where crude oil and natural gas could become in such short supply that burning hydrocarbons as fuel would no longer make sense. In those same circumstances, however, it would still make sense to manufacture petrochemicals because they extract such incredible value from the hydrocarbon molecule.

PETROLEUM MARKETING   14

Crude oil, natural gas, refined products, and petrochemicals are all sold into commodity markets. Commodities are mass–produced, unspecialized products, with high fungibility, having characteristics so similar that they are interchangeable. For example, light sweet crude oil is fungible because a barrel produced in West Texas and one produced in Saudi Arabia would produce similar mixes of products if processed in the same refinery.

Taking a broad view of petroleum marketing, it involves

- Transporting the product to a point where custody transfer is feasible
- Providing storage facilities where necessary
- Balancing production and demand
- Obtaining the best possible price

Crude Oil Marketing

Refineries are normally located near population centers, often in processing clusters that include petrochemical plants. The overall pattern is for crude to be shipped to these clusters from around the world in very large tankers, then to be refined and the products shipped back around the world in smaller tankers.

Since crude is so easily transported by marine tanker, in remote areas of the world the development of a new field generally includes building a pipeline connecting the field to a maring loading terminal. This gives complete access to the world market. An oil company may use its own tankers to ship to its refinery in Europe or the U. S. , but more often it will sell at the terminal to the highly organized and efficient third–party market. Crude oil traders responsible for procuring feedstocks for the world's refineries are in constant contact with the producers, negotiating prices and arranging transportation.

When a new field is discovered in the U. S., the operator normally connects it to the nearest crude oil pipeline and receives the per barrel "posted price" published by the pipeline company. Producers also have the option to rent space in the pipeline to transport their crude. Because of the well established market between producers and refiners in the U. S. and abroad, and because it is a fungible product, no particular attempt is made by companies to run their own crude in their own refineries.

OPEC
(Organization of Petroleum Exporting Countries)

OPEC was formed in 1960 to give the producing countries a unified voice in dealing with the western oil company's crude pricing. Its major impact, however, has been since 1973, when it began functioning as a cartel to control crude prices. In 1999, the OPEC member states are Saudi Arabia, Iraq, Iran, Venezuela, Nigeria, Kuwait, Libya, United Arab Emirates, Indonesia, Algeria, Gabon, and Qatar.

From its beginning, the international crude market was highly volatile. Since major crude oil discoveries were random events that defied planning or scheduling, the supply of crude was usually too long or too short, generating wide swings in price. This environment was discouraging to investment of the huge sums the emerging industry needed. Some stabilization was clearly in everyone's interest.

The first effort to bring order to the market was made by the Texas Railroad Commission in the 1930s, and it was quite successful. At the time, the United States—principally Texas—was the world's biggest crude producer and exporter. The TRC was therefore able to cut back or increase Texas' production as needed to bring world supply and demand into rough balance. Working closely with the major oil companies who controlled the foreign–produced crude, the TRC was able to maintain relative price stability well into the 1960s.

By the early 1970s, the United States had lost its position of leading producer in the world, and in fact had become a net importer. This eliminated the TRC as a player, leaving the multinationals to manage the international crude markets alone. It also gave OPEC new life as it members realized that they now dominated world crude supplies. In 1973, a wave of nationalizations by the producing countries started. OPEC quickly took control of the crude market and in October of 1973 more than tripled prices.

Since then, OPEC has tried to stabilize prices by adjusting member's production rates to meet market demand. Their efforts have been successful at times, but at other times major price fluctuations have occurred. This is not surprising considering the diversity of interests that exist between individual member states. For example, OPEC member Iraq invaded fellow member Kuwait in the Gulf war.

Another factor weakening OPEC is the substantial non–OPEC production that has recently been developed around the world. As of the late 1990s OPEC production has been reduced to only 40% of world production. It is significant, however, that OPEC, particularly Saudi Arabia, still has virtually all the world's surplus producing capacity. This assures them continued influence on crude prices.

Natural Gas Marketing

Because natural gas must be contained under pressure or it will dissipate, commercializing (marketing) of remote gas inevitably involves heavy investment in downstream facilities. One approach is liquefied natural gas (LNG), where the gas is converted in place to a more easily transported liquid. Another approach is to bring demand to the gas field by building, for example, a fertilizer plant in the remote location. The higher valued fertilizer can then be economically transported to the consuming area.

Unlike crude oil, it is not feasible to build surface storage facilities for gas. Gas is, in effect, stored in the reservoir and not produced until it can be consumed. Underground gas storage, usually using depleted gas fields near the market, is a variant of this. In the U. S., the bulk of gas production is in the southwest of the country while the population and consumption is concentrated in the northeast. Demand is heaviest in the winter, so the pipeline operates at maximum rates during cold weather. In the summer, when demand slows, production would be normally be reduced and the pipeline operated at reduced levels. With underground storage the pipeline throughput is kept high. When the gas reaches the northeast, instead of being consumed, it is injected into the subsurface storage reservoir. During the next winter the gas is produced out of storage to augment the gas coming up the pipeline.

With remote gas developments the gas reserves are totally committed to the project, which may last 20 years or longer. This pattern prevails throughout most of the world. In recent years, however, the gas industry in the U. S. has become a great deal more flexible. Deregulation of the industry has freed up pricing, so short term markets for gas, even "spot" markets, have now developed. The contract chain has been altered so that the producer no longer must sell its gas to the pipeline company, who then sells it to the consumer. Instead, the producer and consumer can make the deal direct or through intermediary trading companies, and hire the pipeliner to transport it. This flexibility and efficiency is benefiting all parties. Gas futures are now traded on financial markets, much the same as are crude futures. Companies have become quite sophisticated in placing hedges to cushion themselves from short–term gas price fluctuations.

Refined and Petrochemical Product Marketing

Plant operators not only have to organize and procure their slate of feedstocks, they also have to keep the products they produce moving smoothly out to retail markets. Although temporary storage of excess liquid products is more feasible than storage of natural gas, the available storage capacity is miniscule compared to the enormous volume of petroleum products moving through the system. This smoothly running enterprise again depends on well–developed markets for traders to easily dispose of their surpluses and pick up their shortfalls. As with crude oil, there is no particular attempt made by refiners to move the gasoline they produce through their own service stations only. Instead, a Gulf Coast refiner would typically deliver its surplus gasoline to other marketers in the immediate area in trade for similar volumes received from other refiners on the east coast, west coast, and elsewhere. This saves the cost of physically moving the products around the country, so everyone benefits.

Products are transported from plants by truck, barge, ocean–going ship, and pipeline. A pipeline can handle a variety of different products introduced as sequential batches. Some mixing occurs at the batch interfaces and this material is reprocessed.

LPG is produced not only as a refinery byproduct, but also direct from gas wells. As a byproduct, its production volumes are not determined by its own demand, but rather by the demand for gasoline and natural gas. As a result, the LPG market is notoriously cyclic, swinging rapidly from surplus to deficit and back again. It is therefore necessary to have more LPG storage capacity than for other products. On the U.S. Gulf Coast and elsewhere around the world, this has been accomplished by washing out caverns in subsurface salt formations.

POSTSCRIPT

If this book has done its job, the petroleum industry is now making more sense to you. Whether you used it as an occasional reference or read every page, some of the mists should be cleared away, allowing you to approach your job with more confidence.

If you're interested in additional training, I teach a one-week course using this book as the text. For information, call OGCI Training at (918) 828-2500 or fax them at (918) 828-2580. If you're ready for more specialized training in the various disciplines such as reservoir engineering, seismic exploration, etc., OGCI offers courses and PennWell publishes texts at a non-technical level. For example, Pennwell's Petroleum Refining for the Non-technical Person by William L. Leffler was helpful to me in writing Chapter 12.

— *Chuck Conaway*

GLOSSARY

A

Annulus. In drilling, the space between the drillstring and the hole. Also, the annular space between any two strings of pipe, either casing in casing or tubing in casing.

Anomaly. Petroleum exploration is a process of progressively eliminating the ordinary to bring focus onto the unusual (anomalous) features that might indicate a hydrocarbon accumulation.

Anticline. Upward formation fold. Also called a dome or high.

A.P.I. (American Petroleum Institute) gravity. The density measure used for petroleum liquids. The higher the gravity, the lighter the liquid.

Appraisal well. Well drilled after the field has been discovered to appraise its extent. Particularly used offshore to establish the optimum platform location.

Artificial lift. The use of pumping equipment to lift the fluids out of a well.

Associated gas. Gas that occurs in association with crude oil. Includes solution gas and gas-cap gas.

Automatic custody transfer (ACT). Automated system for selling crude oil at the tank battery.

B

Bent sub. Short joint of drill pipe with a calibrated bend in it. Used in directional drilling.

Bitumen. Naturally occurring near-solid hydrocarbon. Must be mined.

Black oil. Crude oil with strong color (which excludes gas condensate).

Blast joints. Erosion-resistant tubing run in wells to withstand sand impingement.

Blowout. In drilling, when a kick gets out of control and reservoir fluids blow out at the surface.

Blowout preventer (BOP): Safety valve under the rig floor that can be activated to seal off the hole, preventing it from blowing out.

Break out. Unscrew (as in pipe).

Bright spot processing. Seismic processing to detect subsurface gas zones.

Bubble point pressure. When a reservoir is above its bubble point pressure, it has no free gas—all gas is in solution in the oil. As the reservoir is produced and pressure declines, the bubble point pressure is reached. Gas comes out of solution, forming a free gas saturation.

C

Caliper log. Well log to determine hole capacity.

Cap rock. Impermeable rock containing petroleum within trap.

Carbon dioxide flood. Enhanced recovery process where CO_2 gas and water are alternately injected into the formation (WAG process) under pressure to create a miscible flood.

Carried interest. A revenue interest, clear of costs.

Casing. Strings of pipe run in hole and cemented up the backside (annular space between casing and hole).

Catalyst. Compound that assists a chemical reaction but does not participate in it.

Cementation. Cementing together of the grains in clastic rocks.

Chemical weathering. Chemical breakdown of rocks.

Choke. Adjustable orifice on the Christmas tree. Used to control the well's flow rate and protect downstream equipment from full wellhead pressure.

Christmas tree. Valve manifold on top of well.

Clastic rock. Composed of rock grains or fragments.

Closure. The vertical dimension of an oil or gas trap.

Coal-bed methane. Methane recovered by drilling and producing from deep coal beds.

Coiled tubing. Small diameter, continuous (no joints) tubing that is rapidly run in and out of the hole under pressure. Used for many workover and maintenance jobs.

Completion. To make a producing well out of the hole.

Condensate. Produced liquids that were in gaseous form under initial reservoir conditions.

Connate water. Water saturation in rock pores.

Convection. Heat transfer mechanism involving fluid currents driven by heat/density differentials.

Coring. Drilling with a doughnut-shaped bit that allows a cylinder-shaped core of undrilled rock to rise up inside the pipe above the bit. The core is recovered when the drillstring is tripped out of the hole.

Cracking. Breaking up large molecules into smaller ones.

Critical saturations. The minimum saturations of oil, water, or gas in the reservoir that cause the fluid to be a continuous medium, and therefore producible.

Crust. The solid surface of the earth.

Cryogenic. Very low temperature.

Cuttings. Pieces of the drilled rocks brought to the surface by the returning mud stream.

Cyclic steam flooding (huff and puff). Enhanced recovery mechanism for heavy oil. Steam is injected into a well for several days to heat the oil in the surrounding formation and lower its viscosity. The well is then put on pump and the heated oil is produced. The cycle is then repeated.

D

Day-rate drilling contract. Operator pays the contractor a fixed rate for rig time, usually expressed in $/day.

Delta. Area of heavy deposition at the mouth of a river.

Density log. Radioactive well log to determine formation porosity.

Development drilling. Drilling up the field after the exploratory and appraisal wells have been completed.

Diagenesis. The alteration of sediments by the heat, pressure, and chemical conditions encountered during deep burial.

Differential sticking. A differential between the mud column pressure and the reservoir pore pressure forces the drill pipe against the side of the hole and sticks it.

Dim spot processing. Seismic processing to detect "halos" of gas above petroleum traps.

Directional drilling. Drilling at an angle instead of vertically is particularly necessary offshore, where multiple wells need to be drilled from a central platform, but with their bottom-hole locations offset considerably in all directions.

Dogleg. A crooked hole.

Downstream petroleum industry. Transportation, refining and petrochemicals.

Drawworks. On a drilling rig, the large rotating drum that spools the drilling line in and out to raise or lower the load.

Drill collars. Thick-walled joints of drill pipe run at the bottom of the drillstring to put weight on the bit.

Driller. Foreman of the drilling crew and the hands-on operator of the rig.

Drilling break. Abrupt increase in penetration rate that could indicate porosity has been encountered.

Drill report. Summary of previous day's drilling operations delivered early each morning.

Drillship. Ship-shaped, self-propelled, dynamically positioned vessel used to drill in very deep water.

Drill-stem tests. Production tests run on potential pay zones as they are encountered during drilling.

Drillstring. On a drilling rig, the string of pipe extending from the drilling floor to the bottom of the hole. Includes the bit and other bottom-hole tools.

Drilling unit. Consolidation of several small tracts into a single tract large enough to justify the drilling of a well.

Dry hole. An unsuccessful well. Oil or gas may have been encountered, but not in commercial quantities.

Dual completions. Producing two formations in the same well simultaneously but separately.

Dynamically positioning. System to hold a floating vessel on location with computer-controlled thrusters at all four corners.

E

Electric log. Wireline log measuring the formation's electrical resistivity/ conductivity.

Electric submersible pump (ESP): Artificial lift system employing a downhole multi-stage centrifugal pump and motor.

Elevators. In drilling, the device used to connect to and lift the drill string. The elevators are suspended from the hook and latch under the collar of the top joint of pipe.

Emergent coastline. Coastline rising out of the sea.

Enhanced recovery. All the artificial drive mechanisms such as waterflood, CO_2 flood, steam injection, etc.

Erosion. Movement of rock fragments by fluid currents.

Expatriate. A foreign national employed in a country.

Expendable holes. Holes drilled solely to gather information with no intent to ever complete and produce them.

Exploratory well. Well drilled during the exploration phase, as opposed to development wells that are drilled after a discovery has been made.

Farm out. An operating company with a petroleum lease or concession may "farm out" all or a portion of its property to another operator in return for various compensation.

Fault. A fracture of the earth's crust where one side has moved relative to the other.

Feedstock. Raw materials for a manufacturing plant.

Fire flood. Enhanced recovery mechanism for heavy oil reservoirs where air is injected down wells and ignited. The fire front then burns toward producing wells, warming and breaking the viscosity of the oil ahead of it.

Fluids. A substance that flows and yields to any force tending to change its shape. Both liquids and gases are fluids.

Formation. Layer of rock extending over a broad area.

Formation volume factor (ß). Beta is the factor for the volume change undergone by the reservoir fluids when they are produced. In the case of oil, it is the ratio of the space occupied by a barrel of oil at reservoir conditions to the space occupied by a stock tank barrel (STB) of oil. Most oils shrink when their solution gas dissipates at the surface and because of the cooler surface conditions, so their Betas are >1.

Four-dimensional seismic. The fourth dimension is time. A reservoir is re-shot on a regular basis to monitor progress of enhanced recovery project or natural drives.

FPSO (Floating Production, Storage, and Offloading). Vessel used to produce off-shore fields.

Frac-packing. Technique to stabilize unconsolidated formation sand so that it isn't produced.

Fracture porosity. Porosity composed of fractures.

Free-water knockout. Separates free water from emulsion in a tank battery.

Frost wedging. Expanding water-filled fractures in rocks by freezing.

G

Gamma ray log. Well log to determine formation permeability.

Gas cap. The free gas trapped in the top of the structure above the oil leg. When there is a gas cap, the reservoir is at bubble point pressure.

Gas cycling. Technique to increase recovery from retrograde condensate reservoirs. Condensate is stripped out of the production stream and sold, while the residual gas is compressed and injected back into the reservoir.

Gas hydrates. Compounds of gas and water that look like snow or ice but form at higher temperatures. Until natural gas is dehydrated, care must be taken to prevent hydrate plugging in chokes and other cool areas.

Gas lift. Artificial lift system that injects pressurized gas into the well's tubing to lift the oil to surface.

Gas oil ratio (GOR). Ratio of gas to oil produced or in solution in the reservoir. Units are ft³/bbl or m³/bbl.

Gas saturation (Sg). The fraction of the pore volume occupied by gas. Expressed in percent.

Gas separator. Pressure vessel in tank battery that receives the oil, water, and gas from the wells and separates the gas from the oil and water.

GTL (gas-to-liquids) processing. Conversion of natural gas into liquid white products.

Gel. Clay used to build the gel strength in drilling mud necessary to lift drill cuttings out of the hole.

Geologist. Scientist dealing with the earth's processes and rocks. Petroleum geologists focus on finding and recovering oil and gas.

Geophysicist. A scientist specializing in the physics of the earth. In the oil industry, most geophysicists work in seismic exploration, a highly mathematical specialty.

Glacial till. Rock fragments, boulders, and rock dust embedded in glaciers.

Gravel packing. Technique to stabilize unconsolidated formation sand so that it isn't produced.

Gravity platforms. Offshore platforms, typically made of concrete, that sit on bottom and are held in place by their weight, not by piles.

Gravity segregation. Separation of fluids by the force of gravity, which draws the denser material downward, forcing the lighter material upward.

Growth fault. Normal fault along perimeter of depositional basin.

Gunbarrel. See "wash tank".

H

Hard rocks. Well consolidated, harder rocks typically encountered in drilling older, continental sediments. The younger sediments typical of coastal margins are softer.

Heavy oil. Oil with $< 20°$ A.P.I. gravity. Heavy oil typically is viscous and dark colored.

Horizontal drilling. Horizontal directional drilling.

Huff and puff. See Cyclic steam flooding.

Hydraulic fracturing. Stimulation treatment.

Hydrosphere. Water in the oceans, lakes, rivers, etc.

Hydrocarbon molecules. Molecules made up of hydrogen and carbon atoms.

Hydrocarbons. Materials composed of hydrocarbon molecules.

I

Indirect heater. Emulsion treating vessel in a tank battery that circulates emulsion through a tube-bundle immersed in a hot water bath.

Igneous rock. Rock formed by cooling and solidifying of magma.

Infill wells. Wells drilled in between existing wells to reduce well spacing.

Inter-granular porosity. Pores are the spaces between the grains or fragments of clastic rocks.

Irreducible minimum saturation. The water saturation that remains after complete displacement of the pores with oil or gas. This water is affixed to the rock matrix by capillarity, making it essentially immovable.

Isomerization. Rearranging the structure of a molecule without adding or removing any of the atoms.

J

Jacket. Section of offshore platforms extending from the mudline to just above the water line.

Jackknife rig. The usual onshore drilling rig configuration. The derrick "jack-knifes" down, then the rig is broken down into pieces that are transported by truck to the next location.

Jackup rig. Offshore drilling rig that is towed onto location with its legs raised up into the air. At location, its legs are lowered to bottom and the rig is jacked up above wave height. Since they are bottom-supported, jackup rigs are limited to moderate water depths, generally less than 500'.

Jars. Downhole tools that deliver upward hammer blows to dislodge stuck pipe.

Jugs. Onshore seismic microphones.

K

Kelly joint. The top joint in the drillstring. The kelly has flat sides used to transmit rotational force to the drillstring from the rotary system. Not all drilling rigs use a kelly system.

Kick. In drilling, the entry of reservoir fluids into the hole, displacing some of the mud.

Kill a well. Loading the hole with a fluid dense enough to overbalance the formation and prevent flow into the wellbore.

L

Land person. Lawyer or para-legal in United States who represents the operator in negotiating oil and gas leases with mineral owners.

Laterals. Short lateral extensions of a hole drilled with horizontal drilling technology. A well can have multiple laterals, usually intended to increase the well's exposure to the producing formation.

Liquefied natural gas (LNG). Methane liquefied by cryogenic temperatures.

Liquefied petroleum gas (LPG). Petroleum gases (propane and butane) maintained in liquid form by pressure.

Liner. String of casing that does not extend to the surface.

Lithification. Sediments becoming rock.

Lithology. Type of rock.

Lost circulation. In drilling, when drilling mud is lost to a thief zone—often a fractured formation—preventing the continued circulation of mud back to the surface.

LNG. See "liquefied natural gas".

LPG. See "liquefied petroleum gas".

Lubricator. Device that bolts on top of the Christmas tree to run tools in the hole under pressure.

M

Make up. Screw together (as in pipe).

Mantle. Segment of the earth between the core and the crust and the source of magma flows.

Magma. Molten rock material originating within the earth.

Marine. Relating to the sea.

Marker bed. A subsurface horizon with such an abrupt change in acoustic character that it reflects some of the seismic pulse back to the surface. Marker beds show up well on seismic profiles, revealing the structure.

Meander. Bend in a mature, low-gradient river.

Metamorphic rock. Rock altered by heat and/or pressure.

Microbial gas. Natural gas generated by microbial breakdown of vegetable matter.

Migration. Movement of reservoir fluids through a permeable formation, driven by density differences; e.g., oil floats on water.

Miscible flood. Enhanced recovery mechanism where materials are injected into the formation to create miscibility between the formation oil and water, thereby enhancing the effectiveness of water injection.

Moraine. Deposit of glacial till caused by glacial melting.

Mousehole. Hole in the drilling rig floor where a joint of pipe is stored waiting to be added to the string.

Mud motor. Used for directional drilling, the motor is mounted at the bottom of the drillstring just above the bit. Mud pressure rotates the motor, which rotates the bit, without having to rotate the entire drillstring.

N

National oil company (NOC). State-owned companies established to oversee the operations of foreign companies and, in some cases, to independently carry out exploration and production operations.

Natural gas. Naturally occurring hydrocarbon gases, predominantly methane.

Neutron log. Well log measuring formation porosity.

Non-associated gas. Gas from solely gas reservoirs where no black (non-condensate) oil is present.

Normal fault. A tensional fault with movement along the fault plane extending the crust.

O

Offset drilling. When a successful well is drilled close to a lease line, an offsetting well may be drilled on the other side of the line to prevent migration of reservoir fluids across the line.

Oil saturation (So). The fraction of the pore volume occupied by oil. Expressed in percent.

Oil seeps. Where migrating oil emerges at the surface.

Open hole. An uncased well.

Open-hole logs. Well logs run in the open hole before casing is run.

Orifice meter. Determines gas flow rate by measuring the pressure drop incurred by the gas in passing through an orifice.

Outcrop. Rock formation exposed at the earth's surface.

Overbalance. When the pressure generated by the fluid column in the hole is greater than the pore pressure of the rock.

Overburden. Weight of overlying sediments.

Over-riding royalty interest. A fixed share of a royalty interest (e.g. $1/8$th of $7/8$ths).

Ownership in fee. Ownership of both the surface and the mineral rights.

Ox-bow lake. A cut–off meander.

P

P & A (plug and abandon). Permanently plugging a hole with several slugs of cement to prevent fluid movement.

Packer. Downhole tool that seals off the annular space between tubing or drillpipe strings and the hole.

Permeability. The connectivity between the pores of a rock.

Petrochemical. Chemical compound derived from hydrocarbon molecules plus, in some cases, inorganic compounds.

Petroleum. Naturally occurring hydrocarbon compounds.

Petroleum gases. Defined as hydrocarbons that are in a gaseous state at standard conditions of one atmosphere (14.7 psi) and 60° F. Included are methane, ethane, propane, and butane.

Petroleum liquids. Defined as hydrocarbons that are in a liquid state at standard conditions of one atmosphere (14.7 psi) and 60° F. This includes pentanes and heavier.

Petrophysics. Well logging.

Pig. Cleaning or inspection device introduced into a pipeline and propelled down it by the pipeline fluids.

Pipeline proration. Gas-well gas is stored in the reservoir. The wells are opened up when pipeline pressure declines and closed in when pipeline pressure increases.

Plankton. Free-floating organisms.

Plastic. The state of matter with characteristics intermediate between liquids and solids. Plastics flow under pressure (extrusion).

Plate tectonics. Theory visualizing the earth's crust as large moving plates driven by enlargement at spreading centers and reduction at subduction zones.

Polymerization. Combining small molecules into larger ones.

Pooling or unitization. The combining of properties operated by two or more companies into a single property with a single operator. Each of the previous owners have a fixed interest in the pooled operation.

Pore pressure. Pressure of the fluids in the pores of the rock.

Porosity (Φ). The fraction of the rock volume that is pore space. Expressed in percent.

Power swivel. see "Top drive".

Pressure maintenance. Initiation of secondary recovery processes immediately without waiting for primary depletion.

Primary cementing. The initial cementing of casing strings.

Primary recovery. Recovery from the reservoir using only the natural reservoir energy.

Production payments. Pledging a fixed percentage of the revenue stream from specified producing properties a repayment for a loan.

Progressing cavity pump. Artificial lift system using a surface motor to rotate a sucker rod string that drives a downhole pump consisting of a rotating worm-shaped impeller inside of a flexible stator.

Prospect geology. Focusing closely on a small area to develop a drilling recommendation.

Pulling unit. Mobile onshore workover units.

R

Rathole. Hole in the drilling rig floor where the kelly joint and swivel are stored.

Recovery factor. The fraction of a reservoir's original oil in place (OOIP) or original gas in place (OGIP) that will be recovered. Expressed in percent.

Relief well. Well drilled to kill a wild well by penetrating the reservoir close to the wild well and pumping water or cement into it. Relief wells are used only when the wild well cannot be controlled from the surface.

Regional geology. Taking a broad scale view, perhaps looking at an entire basin.

Relative permeability. The degree to which a reservoir's absolute permeability is reduced by the presence of a second fluid; e.g., the permeability to oil is impaired by the connate water, and the higher the Sw, the greater the impairment.

Residue gas. Marketable natural gas remaining after removal of the heavier hydrocarbons, water, and other impurities.

Residual oil or gas saturation. Saturation below which the oil or gas is noncontinuous in the pores, and therefore not producible.

Reserves. Volumes of oil and gas in the reservoir that are commercially producible.

Reservoir. Naturally occurring subsurface deposit of petroleum.

Reservoir Engineer. A specialty of petroleum engineering focused on petroleum reservoir behavior.

Reservoir pressure. The pressure of reservoir fluids.

Reservoir rock. Rock with adequate porosity and permeability to function as a petroleum reservoir.

Resistivity log. Well log measuring the resistivity of formation fluids to determine saturations.

Retrograde condensate reservoir. Gas reservoir where pressure depletion causes condensation of gas liquids in the reservoir.

Reverse fault (also "thrust" fault). Tensional fault with movement along the fault plane shortening the crust.

Rock. Aggregate of mineral grains and crystals.

Roughneck. Member of the drilling crew.

Roundtrip. In drilling, to come out of hole with the drillstring to change a bit or other equipment, then re-run the drillstring to bottom.

Royalty interest. A fixed share of the revenue, clear of costs, from the sale of oil and gas from a property.

S

Salt dome. Upward extrusion of salt from deep salt formations.

S&W. Sediment and water.

Secondary porosity. A second, less significant, porosity system in a rock; e.g., a sandstone's primary porosity may be inter-granular with a secondary fracture porosity.

Secondary recovery. Recovery from a reservoir by adding energy, e. g., water-flooding. Follows primary recovery.

Sedimentary rock. Rock formed from deposited sediments.

Seismic migration. Processing the raw data to correct for tilted formations.

Seismic processing. Processing of raw seismic data to refine it into a useful form. Geophysicists use supercomputers to handle the enormous quantities of data.

Seismic technique. Defining subsurface structures by generating acoustic pulses that penetrate to depth, reflect off subsurface surfaces, return to the surface and are recorded.

Shale shaker. In drilling, device that screens drill cuttings out of drilling mud returns.

Signature bonus. In U.S., cash payment from operator to mineral owners upon their signing oil and gas lease. Usually expressed in $/acre.

Semi-submersible rig. Very large, floating, self-propelled vessel, usually anchored but sometimes dynamically positioned. Drilling position is partially submerged to gain stability.

Sidewall coring. Wireline procedure to get cores from unconsolidated sediments.

Slick-line. Wireline unit using single-strand wire to run through-tubing tools under pressure.

Slip fault. Fault with horizontal movement along fault plane.

Soft rocks. The more recently deposited, unconsolidated, softer rocks typically encountered while drilling coastal margins. The older rocks typical of continental drilling are harder.

Solution gas. Gas that is in solution in the oil under the initial reservoir pressure and temperature. The gas is in the liquid state, not the gaseous state.

Solution gas drive. Reservoir drive mechanism dependent on the expansion and movement of solution gas to move the oil into the wellbore.

Sonic log. Acoustic well log to determine formation porosity.

Sorting. The degree to which the grains are of similar size in a sandstone.

Sour crude. Containing over 2.5% sulfur.

Source rock. Rock containing organic material that generated petroleum.

Spontaneous Potential (SP) log. Well log to determine formation permeability.

Spreading center. Area where new crust is being formed between spreading crustal plates.

Squeeze cementing. Remedial cementing.

Spar. Buoyant offshore platform consisting of a vertical cylinder affixed to bottom and stabilized by anchor lines.

Stabilizer. Large diameter drill collar that "packs the hole" to act as a pivot point in horizontal drilling.

Stand. Three joints of drillpipe, made-up. When tripping out of the hole, drillers do not break out each joint. To save time, they break out every third joint, standing the 90' "stands" of pipe up in the derrick.

Static correction. Seismic processing to adjust for near-surface effects of loose soil, sand dunes, permafrost, etc.

Steam flood. Enhanced recovery mechanism for heavy oil. Steam is introduced through injection wells, then is forced through the formation, heating the oil ahead of it to lower viscosity, which drives the oil to production wells.

Stratigraphic traps. Traps created by depositional-erosional features, not deformation.

Streamer cables. Cables with embedded microphones towed behind marine seismic vessels.

Stimulation treatment. Downhole treatment to increase a well's productivity.

Structure. Distortions of rock formations including folding and faulting.

Structural trap. Traps created by the deformation of rocks, e.g., folding or faulting.

Subduction zone. Area where two crustal plates collide and one overrides the other.

Submergent coastline. Coastline sinking into the sea.

Subsea completions. Offshore wells with Christmas trees at the mudline rather than above the water on a platform.

Sucker-rod pumping unit. Artificial lift system with a surface unit that reciprocates a sucker rod string that strokes a downhole piston-type pump.

Surface geology. Surveying the orientation of rock outcrops to divine subsurface structure.

Suture zone. Area where two crustal plates collide, with neither one overriding the other.

Syncline. Downward formation fold.

Sweep efficiency. Fraction of the reservoir swept by enhanced recovery fluids. Expressed in %.

Sweet crude. Containing 0.5% or less sulfur.

T

Tank battery. The location of central processing, measurement and storage facilities in an oilfield.

Tender-supported rig. The tender is a barge towed into place and anchored beside a permanent production platform. The drilling rig is lifted from the tender to the platform with special hoisting equipment on the tender. Support facilities such as mud pits, generators, crew quarters, cement mixing, pipe storage, etc. remain on the tender, with umbilical lines supplying mud, electric power, etc. to the rig.

Tension-leg platform. Buoyant platform, tethered to bottom by strings of steel pipe under tension.

Tertiary recovery. Enhanced recovery process following secondary recovery. For example, CO_2 flooding after waterflooding.

Thief zone. Zone with anomalously high permeability that drains off a disproportionate amount of injected fluid in an enhanced recovery project or of mud in a drilling well.

Three-phase separator. Separates well fluids into oil, water, and gas streams.

Tight reservoir. Low permeability reservoir.

Time value of money. Perspective focusing on the economic benefits of accelerated revenue from a project. That is, it's better to have $100 in–hand today than to receive $100 a year from now. The $100 in–hand will earn interest during the year and be worth more than the delayed payment.

Toolpusher. The drilling contractor's employee in charge of the rig.

Top drive or power swivel. Surface-located hydraulic motor that rotates the drillstring.

Tour (pronounced "tower"). Drilling rig crew shift. A rig has either three 8-hour tours or two 12-hour tours.

Transform movement. Two crustal plates sliding horizontally past one another with essentially no vertical movement.

Trap. Geologic feature where petroleum could accumulate.

Three-dimensional seismic. 3-D seismic shoots a much denser grid than 2-D by using multiple arrays of microphones off to the side of the shot points. The data is used to build a 3-D mathematical model of the prospect on a minicomputer or workstation. The analyst can then make electronic cuts across the prospect at any angle.

Tubing. Retrievable string of pipe run inside casing to be the conduit for produced fluids.

Turn-key contract. Drilling contract where the contractor is paid a fixed sum to deliver a hole drilled to the operator's specification. Terms are usually expressed in $/ft.

TVD (true vertical depth). The vertical penetration of a well as opposed to its measured depth, which in directional holes can be significantly greater than TVD.

Two-dimensional seismic. The shot points and microphones are on the same line.

Two-phase. Having both a liquid and a gaseous phase; e.g., two-phase pipeline or separator.

U

Underbalance. When the pressure generated by the fluid column in the hole is less than the pore pressure in the rock.

Up-dip pinchout. Stratigraphic trap formed by the lithology changing from permeable to impermeable up-dip, thereby forming a cap rock.

Upstream petroleum industry. Exploration, drilling, and production.

V

Vibroseis unit. An alternative to explosives for generating seismic pulses onshore. A vibroseis truck drops a steel pad from its underbelly, jacks itself up on the pad, then vibrates the pad to generate shock waves.

Virgin oil. Oil in freshly deposited organic material. It has not yet been changed into crude oil by diagenesis—the effects of heat, pressure, and the chemical conditons encountered during deep burial.

VLCC. Very large crude carrier.

Vug. Solution-derived cavity in rock.

Vugular porosity. Porosity composed of vugs.

#

WAG (water & gas) process. Enhanced recovery injection program where slugs of water alternate with slugs of other materials.

Wash tank (or gunbarrel). Large tanks providing retention time for water to settle out of oil-water emulsions.

Water drive. Reservoir drive mechanism dependent on water influx to limit reservoir pressure decline.

Water flood. Enhanced recovery mechanism where water in pumped down injection wells to drive oil through the formation to producing wells.

Water saturation. The fraction of the pore volume occupied by water. Expressed in percent.

Weathering. Breakdown of rock by physical and chemical effects.

Well control. In drilling, handling kicks and preventing blowouts.

Wellhead. Located under the Christmas tree, it is the landing device for casing and tubing strings. Also, a general term used for the well's surface equipment, including the tree, e.g., "wellhead pressure".

Well log. Measurement of downhole rock characteristics. Logging tools are run to bottom on wireline, then drawn slowly upward through the zones of interest, recording signals emitted spontaneously by the formations or induced by emissions from the logging tool.

White products. Gasoline, diesel, kerosene, and heating oil.

Wild well. Well that has blown out of control.

Wildcat well. Exploratory well drilled some distance from existing producing wells.

Workover. Remedial downhole work on a well.

INDEX

blending, 240–241
Refining and petrochemicals, 231–243
 feedstocks, 231–232
 crude oil transportation, 232–233
 refinery operations, 233–241
 petrochemicals, 241–243
Regasification terminal, 228
Regulatory functions, 63
Regulatory oversight, 60–63
 protecting landowner's rights, 60–61
 maximizing efficient recovery,
  61–63
Relative permeability, 29–30
Relief well, 114
Repairing casing leaks, 183
Replacing well equipment, 183
Reserve determination methods,
 92–95
 volumetric method, 93
 material balance method, 93–94
 reservoir modeling/simulation, 94
 production decline curve method,
  94–95
Reserves (oil and gas), 91–95
 categories, 92
 reserve determination methods,
  92–95
 reserve revisions, 95
 reserves not assets, 95
Reservoir fluids, 69–73, 145
 definitions, 69–71
 fluid systems, 71–73
 samples, 145
Reservoir friction, 186–187
Reservoir heterogeneity, 83–84
Reservoir modeling/simulation, 94
Reservoir performance, 69–95
 reservoir fluids, 69–73

oil reservoirs (primary drives),
 74–79
gas reservoirs, 79–80
waterflooding, 80–85
thermal recovery, 85–88
miscible flooding, 88–90
mobility ratio improvement, 90
microbial floods, 91
oil and gas reserves, 91–95
Reservoir pressure, 37–39
 cause, 37–38
 abnormal pressures, 38–39
Reservoir rock properties, 24–30
 porosity, 25–28
 permeability, 28–30
 fluid saturations, 29–30
 relative permeability, 29–30
Reservoir rock, 24–30, 33
 properties, 24–30
Resistivity logs, 143–144
Re–stimulation, 183
Retarders, 154
Retention time, 209
Retrievable/temporary packers, 168
Retrograde condensate gas, 73, 79–80
 reservoir, 79–80
Reverse faults, 12–13
Ridge and valley areas, 13–14
Risk/cost (exploration), 41
River deltas, 18
Rivers, 16
Rock properties (reservoir), 24–30
Rock types, 7–12
 sedimentary, 7–10
 formations, 11–12
Rockefeller, John D., xiv–xvi
Rotary drilling system, 98–108
 drillbits, 98–101